WITHDRAWN
UTSA LIBRARIES

RENEWALS 458-4574

Production Testing of RF and System-on-a-Chip Devices for Wireless Communications

For a listing of recent titles in the *Artech House Microwave Library,* turn to the back of this book.

Production Testing of RF and System-on-a-Chip Devices for Wireless Communications

Keith B. Schaub

Joe Kelly

Artech House, Inc.
Boston • London
www.artechhouse.com

Library of Congress Cataloging-in-Publication Data

A catalog record for this book is available from the U.S. Library of Congress.

British Library Cataloguing in Publication Data
Schaub, Keith B.
  Production testing of RF and system-on-a-chip devices for wireless communications.
—(Artech House microwave library)
  1. Semiconductors—Testing 2. Wireless communication systems—Equipment and supplies—Testing
  I. Title II. Kelly, Joe
  621.3'84134'0287

  ISBN 1-58053-692-1

Cover design by Gary Ragaglia

© 2004 ARTECH HOUSE, INC.
685 Canton Street
Norwood, MA 02062

All rights reserved. Printed and bound in the United States of America. No part of this book may be reproduced or utilized in any form or by any means, electronic or mechanical, including photocopying, recording, or by any information storage and retrieval system, without permission in writing from the publisher.
  All terms mentioned in this book that are known to be trademarks or service marks have been appropriately capitalized. Artech House cannot attest to the accuracy of this information. Use of a term in this book should not be regarded as affecting the validity of any trademark or service mark.

International Standard Book Number: 1-58053-692-1

10 9 8 7 6 5 4 3 2 1

To my loving mother, Billie Meaux, and patient father, Leslie Meaux, for being the best parents that any son could wish for and without whose love and guidance I would be lost.

To my beautiful, loving sister, Jamie Ormsbee Cooper, who defines strength and determination and always believed in me.

To my best and most trusted friend, brother, and advisor, John Anthony Bartula, who keeps me grounded and whose friendship I depend on more than words can describe.

—Keith B. Schaub

To my parents, Joseph and Barbara, who helped me appreciate all that life hands to me.

To my grandparents, especially my grandfather, Emil Gasior, who made me realize that anything can be achieved if you put forth the effort.

—Joe Kelly

Contents

Preface — xiii
Acknowledgments — xvii

CHAPTER 1
An Introduction to Production Testing — 1

1.1 Introduction — 1
1.2 Characterization Versus Production Testing — 1
1.3 The Test Program — 2
1.4 Production-Test Equipment — 2
1.5 Rack and Stack — 2
1.6 Automated Test Equipment — 3
1.7 Interfacing with the Test Equipment — 3
    1.7.1 Handlers — 3
    1.7.2 Load Boards — 5
    1.7.3 Contactor Sockets — 5
    1.7.4 Production RF and SOC Wafer Probing — 6
1.8 Calibration — 9
1.9 The Test Floor and Test Cell — 10
1.10 Test Houses — 10
1.11 Accuracy, Repeatability, and Correlation — 10
1.12 Design for Testing — 11
1.13 Built-in Self-Test — 11
    References — 12

CHAPTER 2
RF and SOC Devices — 13

2.1 Introduction — 13
2.2 RF Low Noise Amplifier — 15
2.3 RF Power Amplifier — 15
2.4 RF Mixer — 16
2.5 RF Switch — 19
2.6 Variable Gain Amplifier — 20
2.7 Modulator — 22
2.8 Demodulator — 23
2.9 Transmitter — 24
2.10 Receiver — 24
2.11 Transceiver — 25

2.12		Wireless Radio Architectures	26
2.13		Superheterodyne Wireless Radio	26
2.14		Zero Intermediate Frequency Wireless Radio	26
2.15		Phase Locked Loop	28
2.16		RF and SOC Device Tests	30
		References	31

CHAPTER 3
Cost of Test — 33

3.1	Introduction	33
3.2	Wafer Processing Improves Cost of Test	33
3.3	Early Testing of the SOC	36
3.4	SCM and IDM	37
3.5	SOC Cost-of-Test Paradigm Shift	37
3.6	Key Cost-of-Test Modeling Parameters	38
	3.6.1 Fixed Cost	39
	3.6.2 Recurring Cost	39
	3.6.3 Lifetime	40
	3.6.4 Utilization	40
	3.6.5 Yield	41
	3.6.6 Accuracy as It Relates to Yield	42
3.7	Other Factors Influencing COT	45
	3.7.1 Multisite and Parallel Testing	45
	3.7.2 Test Engineer Skill	46
3.8	Summary	46
	References	46

CHAPTER 4
Production Testing of RF Devices — 49

4.1	Introduction	49
4.2	Measuring Voltage Versus Measuring Power	49
4.3	Transmission Line Theory Versus Lumped-Element Analysis	50
4.4	The History of Power Measurements	51
4.5	The Importance of Power	52
4.6	Power Measurement Units and Definitions	53
4.7	The Decibel	53
4.8	Power Expressed in dBm	54
4.9	Power	54
4.10	Average Power	55
4.11	Pulse Power	56
4.12	Modulated Power	56
4.13	RMS Power	57
4.14	Gain	58
	4.14.1 Gain Measurements of Wireless SOC Devices	60
4.15	Gain Flatness	61
	4.15.1 Measuring Gain Flatness	63
	4.15.2 Automatic Gain Control Flatness	65

4.16	Power-Added Efficiency	67
4.17	Transfer Function for RF Devices	68
4.18	Power Compression	69
4.19	Mixer Conversion Compression	72
4.20	Harmonic and Intermodulation Distortion	72
	4.20.1 Harmonic Distortion	73
	4.20.2 Intermodulation Distortion	75
	4.20.3 Receiver Architecture Considerations for Intermodulation Products	79
4.21	Adjacent Channel Power Ratio	79
	4.21.1 The Basics of CDMA	79
	4.21.2 Measuring ACPR	81
4.22	Filter Testing	82
4.23	S-Parameters	84
	4.23.1 Introduction	84
	4.23.2 How It Is Done	84
	4.23.3 S-Parameters of a Two-Port Device	85
	4.23.4 Scalar Measurements Related to S-Parameters	86
	4.23.5 S-Parameters Versus Transfer Function	88
	4.23.6 How to Realize S-Parameter Measurements	89
	4.23.7 Characteristics of a Bridge	89
	4.23.8 Characteristics of a Coupler	90
4.24	Summary	91
	References	92
	Appendix 4A: VSWR, Return Loss, and Reflection Coefficient	93

CHAPTER 5
Production Testing of SOC Devices 95

5.1	Introduction	95
5.2	SOC Integration Levels	96
5.3	Origins of Bluetooth	97
5.4	Introduction to Bluetooth	98
5.5	Frequency Hopping	99
5.6	Bluetooth Modulation	100
5.7	Bluetooth Data Rates and Data Packets	100
5.8	Adaptive Power Control	102
5.9	The Parts of a Bluetooth Radio	102
5.10	Phase Locked Loop	103
5.11	Divider	104
5.12	Phase Detector, Charge Pumps, and LPF	104
5.13	Voltage Controlled Oscillator	104
5.14	How Does a PLL Work?	104
5.15	Synthesizer Settling Time	105
5.16	Testing Synthesizer Settling Time in Production	106
5.17	Power Versus Time	106
5.18	Differential Phase Versus Time	110
5.19	Digital Control of an SOC	112

5.20	Transmitter Tests	113
	5.20.1 Transmit Output Spectrum	114
	5.20.2 Modulation Characteristics	117
	5.20.3 Initial Carrier Frequency Tolerance	118
	5.20.4 Carrier Frequency Drift	119
	5.20.5 VCO Drift	120
	5.20.6 Frequency Pulling and Pushing	120
5.21	Receiver Tests	124
	5.21.1 Bit Error Rate	125
	5.21.2 Bit Error Rate Methods	127
	5.21.3 Programmable Delay Line Method (XOR Method)	127
	5.21.4 Field Programmable Gate Array Method	128
	5.21.5 BER Testing with a Digital Pin	128
	5.21.6 BER Measurement with a Digitizer	130
5.22	BER Receiver Measurements	132
	5.22.1 Sensitivity BER Test	132
	5.22.2 Carrier-to-Interference BER Tests	133
	5.22.3 Cochannel Interference BER Tests	133
	5.22.4 Adjacent Channel Interference BER Tests	133
	5.22.5 Inband and Out-of-Band Blocking BER Tests	135
	5.22.6 Intermodulation Interference BER Tests	135
	5.22.7 Maximum Input Power Level BER Test	137
5.23	EVM Introduction	137
	5.23.1 I/Q Diagrams	137
	5.23.2 Definition of Error Vector Magnitude	138
	5.23.3 Making the Measurement	139
	5.23.4 Related Signal Quality Measurements	141
	5.23.5 Comparison of EVM with More Traditional Methods of Testing	142
	5.23.6 Should EVM Be Used for Production Testing?	142
	References	143

CHAPTER 6
Fundamentals of Analog and Mixed-Signal Testing — 145

6.1	Introduction	145
6.2	Sampling Basics and Conventions	145
	6.2.1 DC Offsets and Peak-to-Peak Input Voltages	146
6.3	The Fourier Transform and the FFT	147
	6.3.1 The Fourier Series	147
	6.3.2 The Fourier Transform	147
	6.3.3 The Discrete Fourier Transform	149
	6.3.4 The Fast Fourier Transform	150
6.4	Time-Domain and Frequency-Domain Description and Dependencies	150
	6.4.1 Negative Frequency	150
	6.4.2 Convolution	151
	6.4.3 Frequency- and Time-Domain Transformations	152

6.5	Nyquist Sampling Theory	154
6.6	Dynamic Measurements	156
	6.6.1 Coherent Sampling and Windowing	156
	6.6.2 SNR for AWGs and Digitizers	159
	6.6.3 SINAD and Harm Distortion	160
6.7	Static Measurements	163
	6.7.1 DC Offset	163
	6.7.2 INL/DNL for AWGs and Digitizers	164
6.8	Real Signals and Their Representations	165
	6.8.1 Differences Between V, W, dB, dBc, dBV, and dBm	165
	6.8.2 Transformation Formulas	166
6.9	ENOB and Noise Floor: Similarities and Differences	167
6.10	Phase Noise and Jitter	167
	6.10.1 Phase Noise and How It Relates to RF Systems	168
	6.10.2 Jitter and How It Affects Sampling	168
6.11	I/Q Modulation and Complex FFTs	168
	6.11.1 System Considerations for Accurate I/Q Characterization	168
	6.11.2 Amplitude and Phase Balance Using Complex FFTs	169
6.12	ZIF Receivers and DC Offsets	171
	6.12.1 System Gain with Dissimilar Input and Output Impedances	171
6.13	Summary	172
	References	173

CHAPTER 7
Moving Beyond Production Testing — 175

7.1	Introduction	175
7.2	Parallel Testing of Digital and Mixed-Signal Devices	175
7.3	Parallel Testing of RF Devices	175
7.4	Parallel Testing of RF SOC Devices	178
7.5	True Parallel RF Testing	179
7.6	Pseudoparallel RF Testing	180
7.7	Alternative Parallel RF Testing Methods	182
7.8	Guidelines for Choosing an RF Testing Method	184
7.9	Interleaving Technique	185
7.10	DSP Threading	186
7.11	True Parallel RF Testing Cost-of-Test Advantages and Disadvantages	187
7.12	Pseudoparallel RF Testing Cost-of-Test Advantages and Disadvantages	188
7.13	Introduction to Concurrent Testing	189
7.14	Design for Test	190
7.15	Summary	191
	References	192

CHAPTER 8
Production Noise Measurements — 193

8.1	Introduction to Noise		193
	8.1.1	Power Spectral Density	193
	8.1.2	Types of Noise	194
	8.1.3	Noise Floor	198
8.2	Noise Figure		199
	8.2.1	Noise-Figure Definition	199
	8.2.2	Noise Power Density	201
	8.2.3	Noise Sources	202
	8.2.4	Noise Temperature and Effective Noise Temperature	202
	8.2.5	Excess Noise Ratio	203
	8.2.6	Y-Factor	204
	8.2.7	Mathematically Calculating Noise Figure	204
	8.2.8	Measuring Noise Figure	205
	8.2.9	Noise-Figure Measurements on Frequency Translating Devices	209
	8.2.10	Calculating Error in Noise-Figure Measurements	210
	8.2.11	Equipment Error	211
	8.2.12	Mismatch Error	211
	8.2.13	Production-Test Fixturing	212
	8.2.14	External Interfering Signals	212
	8.2.15	Averaging and Bandwidth Considerations	212
8.3	Phase Noise		213
	8.3.1	Introduction	213
	8.3.2	Phase-Noise Definition	214
	8.3.3	Spectral Density–Based Definition of Phase Noise	216
	8.3.4	Phase Jitter	216
	8.3.5	Thermal Effects on Phase Noise	217
	8.3.6	Low-Power Phase-Noise Measurement	217
	8.3.7	High-Power Phase-Noise Measurement	217
	8.3.8	Trade-offs When Making Phase-Noise Measurements	217
	8.3.9	Making Phase-Noise Measurements	218
	8.3.10	Measuring Phase Noise with a Spectrum Analyzer	220
	8.3.11	Phase-Noise Measurement Example	220
	8.3.12	Phase Noise of Fast-Switching RF Signal Sources	222
	References		222

Appendix A: Power and Voltage Conversions	225
Appendix B: RF Coaxial Connectors	229
List of Acronyms and Abbreviations	233
List of Numerical Prefixes	237
About the Authors	239
Index	241

Preface

It came to our attention that there were not any books available that enlightened the engineer on the concepts of production testing of radio frequency (RF) and system-on-a-chip (SOC) devices. There is a number of great books and application notes on the subject of RF measurement techniques. There is also a number of great mixed-signal analysis and measurements how-to books. However, there are no books that bring the two worlds of RF and mixed-signal testing into one volume. It is our intention to bridge this gap.

Under the topic of electronics there are two major categories of devices, digital and analog. Digital refers to those devices that manipulate data between two states (i.e., 1 or 0). Analog refers to the manipulation of continuous waveforms. Analog electronics is a very general topic, and for the most part, the subject falls under the category of mixed signal. Analog measurements are also covered in this category. However, when discussing RF electronics (also analog), special attention must be paid to the rules introduced under the category of mixed-signal testing. It is these rules that often make people approach RF with trepidation. But, they are simply that, rules. If they are followed, RF is very straightforward. An RF engineer could reference back to old college books as these topics and test concepts are derived from the fundamental theories of physics. However, our goal is to present these measurements within this book in a straightforward manner, with explanations covering the gotchas that all of us have run into over time.

Indeed, many of the descriptions will be based on microwave theory and the theory of microwave devices. But this is a necessary foundation, so that topics may be taken two steps further:

1. Describing the test;
2. Explaining how to implement production-testing solutions.

Testing and measuring RF and SOC devices is routinely performed on bench tops in laboratories, but production testing adds the constraints of performing these tests significantly more efficiently, while maintaining the same level of quality. The term *efficiently* commonly means "more quickly," but it can also mean introducing creative means such as multisite testing or parallel testing. Topics such as these will be covered throughout the chapters in this book.

This book is intended for a wide variety of audiences. They include SOC applications engineers, engineering managers, product engineers, and students, although other disciplines can benefit as well. The book is constructed in two parts. The first part consists of the first three chapters, readable like a novel, informing the reader of the details of production testing and presenting items to consider such as cost of test

(COT). The second half (Chapters 4 through 8) is written as a handbook, specifically for applications engineers. It is our intention to create a book that will be used as a reference, providing algorithms and good-practice techniques. Additionally, the appendixes that we have included contain items that would typically be needed by SOC engineers. The book is also aimed at managers of technical teams, that they may pick up this book, read the first few chapters, and feel comfortable in relatively detailed discussions involving applications and production-test solutions.

A few years ago, an RF applications engineer would be very focused in this very unique (often termed *complex*) field performing tests on discrete RF devices such as mixers, power amplifiers, low noise amplifiers, and RF switches. Times have changed. Today, we face increasing levels of integration, such that many of these discrete device functions are contained within one chip or module. Furthermore, the integration levels are such that RF chips contain lower-frequency analog functionality, as well as digital functionality (earlier RF devices often contained three-wire serial communications for controlling things such as gain control, but current digital is becoming more complex). Indeed, it would be more accurate, when referring to this new breed of engineers, to coin the term *SOC engineer* when discussing today's wireless applications.

Chapter 1 provides an overview of the many facets of production testing, with particular focus on the testing of RF and SOC devices. Many of the topics also directly work for other types of electronic device production testing. Additionally, the various capital expense items are covered, such as handlers, wafer probers, load boards, contactors, and so forth. There are not many general information application notes available on these topics, and this chapter is intended to bring them together to one location.

Chapter 2 introduces the devices, both RF and SOC, that this book focuses on. A review of how the radio has evolved in wireless communications is presented. The superheterodyne radio and direct conversion (zero-if) architectures are discussed, as are their changes over time and their impact on testing. Lastly, an overview of the types of tests that are performed on each type of device is presented.

Cost of test is reviewed in Chapter 3. An in-depth analysis is presented in this chapter with the intention to be a guide for those making decisions on how to implement final tests of devices. Note that this chapter, while presented in a book on RF testing, can be applied equally to any other type of electronic device or wafer testing. The intention is for this chapter to be useful to managers, sales teams, and applications engineers who go beyond the role of sitting behind the tester. Also presented in this chapter is a discussion of the traditional models of production test. Topics considered include the advantages and disadvantages of using third-party-testing integrated design manufacturers (IDMs) versus subcontract manufacturers (SCMs). An analytical tool will be presented for calculating cost of test, including many necessary components that are often overlooked when deciding how to perform production testing.

Algorithms for production tests performed on discrete RF devices, as well as the front end of more highly integrated devices are presented in Chapter 4, the beginning of the handbook-type portion of this book. Detailed descriptions of the tests, as well as algorithms in both tabular and block diagram formats, are provided.

Following the format of Chapter 4, Chapter 5 provides algorithms on measurements used with more highly integrated SOC devices. The tests discussed in this section are typical of those found in wireless communications.

Chapter 6 is an introduction to many facets of mixed-signal testing. Common tests that are finding their way into SOC device production testing are explained.

Chapter 7 covers new methods for improving the efficiency of production testing, taking it beyond simply performing the measurements faster. Concepts such as parallel and concurrent testing are presented.

Chapter 8 is dedicated to the measurement of noise. Both noise figure and phase noise measurements are discussed. The intention of this chapter is to educate the engineer in what goes on behind the scenes of today's easy-to-use noise figure analyzers and automated test equipment (ATE). Gone are the days when the engineer had to manually extract noise measurements, but it is important to understand the algorithms, which even today, within analyzers, effectively remain unchanged. There is further explanation on how to perform noise measurements in a production-test environment. Phase noise is also be considered and examined.

Appendixes 4A, A, and B are included to cover the common items that every engineer is often running hastily to find from their notes.

We look forward to helping to merge the worlds of RF and mixed-signal production testing.

Keith Schaub
Joe Kelly
March 2004

Acknowledgments

Keith Schaub would like to thank his coauthor Joe Kelly for agreeing to take on the challenge of writing this book together. Additionally, the authors would like to thank their new friends and coworkers Edwin Lowery III and Ashish Desai. Edwin Lowery wrote all of Chapter 6 on mixed signal testing as it applies to wireless, under an impossible schedule and delivered exceptional quality. Ashish Desai single-handedly wrote about the state-of-the-art topic of error vector magnitude (EVM) and this book has benefitted considerably from his contributions.

The experience that we have gained over the years that has afforded us the opportunity to develop this book is due in part to many of the outstanding engineers that we have had the good fortune to work with throughout our careers including, but certainly not limited to (in alphabetical order):

Advantest
    Donald Cooper
Agilent Technologies, Inc.
    Robert Bartz, Don Blair, Jeff Brenner, Scott Chesnut, Eric Chiu, Bob Cianci, Bill Clark, Peter Eitner, Michael Engelhardt, Frank Goh, Troy Heistand, Ron Hubscher, Miklos Kara, Peggy Kelley, Hiroshi Kikuyama, Ginny Ko, Doug Lash, Anthony Lum, Roger McAleenan, Dan McLaughlin, John McLaughlin, Mike Millhaem, Mike Moodhead, Satoshi Nomura, Laurent Ollivier, Nick Onodera, Darrin Rath, Ted Sato, Jason Smith, Eng Keong Tan, Kim Tran, and Juergen Wolf
DSP Group
    Behrouz Halliyal
Epcos
    Mike Alferman, Ulrich Bauernschmitt, Stefan Freisleben, Joachim Gerster, and Wolfgang Till
Filtronics
    Nigel Cameron
IBM
    Ernst Bohne and Angelo Moore
Infineon
    Klaus Dahlfeld
Globespan
    Mark Wilbur
Motorola
    Doug Jones, Erica Miller, and Kern Pitts

Philips
    Mike Bellanger
Qualcomm
    Farzin Fallah and Pat Sumner
RF Micro Devices
    Igor Emelianoff
Rutgers University
    Ahmad Safari, Yicheng Lu, Sigrid McAfee, and Daniel Shanefield
Schlumberger
    Rudy Garcia
Silicon Wave
    Brian Pugh and Phong Van Pham
Texas Instruments
    Carsten Schmidt and Friedrich Taenzler
U.S. Army Research Lab
    Arthur Ballato and John Vig

CHAPTER 1
An Introduction to Production Testing

1.1 Introduction

For many years, radio frequency (RF) devices have been tested only to ensure that they perform to specifications. Up until the early 1980s there were not many wireless consumer devices. Most wireless devices at the time were used in military applications. The tests performed on these devices were long and time-consuming to assure near-perfect operation in radar-based applications or their other intended purposes.

In the later 1980s the pager was introduced. Consisting of simply a receiver, this was the beginning of the need for testing of RF devices in large volumes. In the early 1990s RF technology emerged into the consumer market in the form of cordless and wireless (cellular, mobile) phones. There was a subsequent market explosion and an immediate proliferation of mobile phones. It was apparent that the industry had expanded and as a result the prices of semiconductor devices dropped significantly, especially when compared to the RF devices used for military applications.

Now, as it is critical to produce quality and properly working products, RF and system-on-a-chip (SOC) semiconductor devices are tested 100% for their intended functionality. The difficult task is to derive a means to provide an efficient and comprehensive test methodology that can accurately sort good parts from defective parts, and at a low cost. As will be discussed in Chapter 3, the cost of test of modern RF and SOC devices has become a significant part of the overall cost of producing these devices.

Therefore, production testing of RF and SOC devices is the act of performing numerous tests in a short amount of time on high volumes of parts. The major objective is to have high throughput and low overhead, or low cost of test, such that the production testing does not adversely impact the marketable value of the device.

1.2 Characterization Versus Production Testing

Testing of a device under test (DUT[1]) can be performed in a number of ways. In production testing, it is optimal to have the shortest test needed to pass good DUTs and fail bad DUTs. When a test program reaches the full production-testing stage, there should be a minimal number of tests utilized. In contrast, during the early stages of production and preproduction runs, the test program is often conservatively

---

1. The term UUT, for unit under test, is a more general production-testing terminology that is sometimes used when discussing testing of electronic devices.

written, so that the DUT is overtested (redundant test coverage). This is attributed to the number of people involved in the development of the device, each with a specified set of tests to satisfy individual criteria. This methodology may initially ensure designer confidence, but as the test program matures (usually over a period of many weeks), tests are removed; thus, the final production-test program may not even resemble the initial test plan.

There are additional reasons for a large number of tests in a test program. In early stages of the product life cycle, the design and manufacturing engineers of the DUT seek awareness of potential production flaws and tolerances. This is best achieved by feeding back excessive quantities of information from the tests. Even as the product matures and the test list is reduced, a test program may include provisions to run extensive tests on every nth part. This is known as characterization test.

1.3 The Test Program

A test program (also called test plan or test flow) is a computer program that tells the test system how to configure its hardware to make the needed measurements. This program can range from low-level C/C++ code to a graphical interface for ease of use. Within this program, instructions to the hardware and information such as how to determine if the DUT has passed or failed the test (known as limits) is provided.

1.4 Production-Test Equipment

From the moment an RF or SOC device has been fabricated on a wafer or placed into a package, testing of the device occurs in a laboratory environment. The test equipment used may range from simple multimeters to network analyzers. If there is a number of different tests to be performed routinely on a device, then often an engineer will group equipment in a common locale for the convenience of being able to perform all the measurements with ease.

This model defines a rudimentary test system, as it has all of the equipment in one location to perform all necessary tests. However, it is not yet production worthy as defined. A production-test system, or tester, also has the means to quickly place a DUT into and out of the test setup and virtually eliminate human interaction when testing a large number or group of parts.

Production-test equipment comes in two primary architectures: rack-and-stack assemblies and automated test equipment (ATE) configurations. Characteristics of both are discussed along with their advantages and disadvantages to ease the selection of the appropriate solution.

1.5 Rack and Stack

Similar to the laboratory configuration mentioned above is the rack-and-stack configured tester. This is a suitable configuration for a production tester during the characterization and prototype stages of a device because the equipment contained on the rack can be quickly reconfigured to meet changing needs. Often rack-and-stack configurations are customized to a specific part. This is an advantage and a disadvantage. The custom tailoring is advantageous as it can enable the fastest possible test times. It can also be a disadvantage in that it reduces the flexibility of the architecture. Often, the tester has to be significantly rebuilt for another product to be tested. The computer programs that run the hardware can also be somewhat difficult as there may be interfacing with the equipment via various buses or protocols.

1.6 Automated Test Equipment

Automated test equipment (ATE) is a tester that is designed as a complete stand-alone unit for optimal production testing of devices. This is the primary advantage of ATE. Many of the larger test-equipment manufacturers produce these systems. Optimally designed systems are flexible and, with respect to RF and SOC devices, can also test a multitude of parts. The manufacturers of ATE consider market factors when designing testers of this type. They focus on usability and flexibility in architecture and ease of programming for the user.

1.7 Interfacing with the Test Equipment

Once the test equipment is established, an efficient means to route the signals from the test equipment to the DUT must be determined. Many pieces fit into this puzzle, such as load boards, contactors, handlers, wafer probes, wafer probers, and the like. The following sections describe these key items.

1.7.1 Handlers

When production testing of any packaged semiconductor device is performed, one of the major capital investments is the handler. The handler is a robotic tool for placing the DUT into position to be tested. The foremost determinant of the type of handler is based upon how the devices are delivered to the final production testing stage (i.e., trays, tubes, and so forth). After the test is performed, the handler then places the DUT into an appropriately selected pass bin or fail bin as determined by the tester. Handlers are found in many varieties and have many different features. This section will provide an overview of handlers, which includes information critical for the handler selection process. In searching, we have found little documentation on the overview of handlers for production testing, but references at the end of this section can provide more detailed information on the specific types of handlers.

First and foremost, handlers come in as many varieties as package types. The two major handler types are gravity feed and pick and place.

Gravity feed handlers work best for packages that are mechanically quite solid and can withstand friction on a sliding surface, such as the following package types: small outline integrated circuit (SOIC), miniature small outline package (MSOP), thin small outline package (TSOP), and leadless chip carrier (LCC). A gravity feed handler has the DUTs usually fed into a slider via transportation tubes. When the

DUT gets to the slider, it slides down to the load board due to gravitational force. Because smaller, lighter packages pose a problem with friction, some handlers integrate air blowers into the channel along the gravity slider to assist in the acceleration of the DUT to the load board.

Pick-and-place handlers can work with almost all type of packages. Typically using suction, this handler moves the DUT from a transportation tray to the load board contactor socket. The precision movement in these handlers is controlled through stepper motors. Pick-and-place handlers often employ numerous vacuum solenoids, rather than electrically controlled switches, which minimize the introduction of noise to the production-testing environment.

Index time, or the time that it takes to place a tested DUT into the appropriate bin and obtain and place a new DUT into the contactor socket, can be a critical factor, especially when the test-plan execution times are less than a second. Typical handler index times range from 0.4 to 0.75 seconds. For example, if the time to execute an entire test plan takes 0.5 second and the index time of the handler is 0.5 second, it is clear that only half of the processing time is actual testing. This demonstrates the benefit of multisite testing, which, in addition to being dependent on the tester software, is also highly dependent on the handler configuration and capabilities. Additionally, on the topic of index time, it is recommended to place the most highly accessed bins closest to the contactor socket so that the mechanical motion of the handler is minimized, thereby reducing index time. For example, if the yield of a given lot is 80%, then it would be beneficial to place the "good," or "pass," bins nearest to the contactor socket. This would enable the shortest range of motion for the most common task. Gravity feed handlers typically have shorter (better) index times than pick-and-place handlers.

The number of sites that a handler is capable of providing is also important. The number of sites available on a handler can be anywhere from 1 to more than 32 sites. However, for RF/SOC testing, quad-site is considered the state-of-the-art method. Handlers with more than four sites are designed to accommodate devices with a high degree of digital testing or built in self-testing (BIST), such as memory devices.

Additionally, handlers may have to be used for environmental testing, such as testing the DUT across various temperature ranges. When operating a handler under thermal conditions, a handler may need to provide cooling as well as heating capability. Typical ranges are from –60°C to 160°C. Another feature that may be necessary is thermal soaking, or maintaining the DUT at a set temperature prior to or during testing. Conventional means of providing an environmental temperature are through the use of liquid nitrogen or chilled water. Other technologies for cooling and heating are forced-air cooling or coolant mixing.

The size of the handler, or its footprint, may or may not be a significant factor in the decision of which handler to use. It is important to note that with capital equipment, floor space is money. To allow the reduction of floor space required for production testing and to eliminate excess time, additional functionalities can be integrated into some handlers, such as DUT lead inspection and placement into tape and reel for shipping.

Autoloaders and unloaders of trays and tubes, or any other means in which the DUTs are delivered to the production-testing stage, make the testing process much easier. Requiring a handler operator to load and unload DUTs into a handler leads to a significant decrease in yield. This comment is from first-hand experience; for example, conversations between test floor operators about social events from the previous evening often take precedence over the empty device feed in the handler.

1.7.2 Load Boards

A load board is defined as a printed circuit-board assembly that is used to route all of the tester resources to a central point that then allows the DUT to perform during its test time. This assembly may also be referred to as a DUT interface board (DIB).

The load board is independent of the tester and is almost always unique to each DUT that is tested. One of the most time-consuming elements of developing a full production-test solution is the design and fabrication of the load board. It must be considered that all of the dc power supply, digital control, mixed signal, and RF signal lines must coexist and be routed among each other on a common board. This inevitably requires a multilayered load board to be fabricated. Creating a load board is a process, including design, layout, fabrication, assembly and test, and possibly multiple redesigns. The making of the load board is very similar to the fabrication of the actual DUT, although not as complicated, and ample time for this effort should be included in the project schedule.

Another often-overlooked difficulty is the final impedance matching and tuning that is necessary after the board is fabricated. Time should be allowed for this effort, especially if it is being done for the first time. Having an experienced RF circuit tuning person on the team would help save significant time in this area. Alternatively, close communication with the DUT designer can provide time-saving tips, as he or she would be aware of areas of the device that are sensitive to impedance matching.

Additionally, there are many third-party companies that provide services from consulting to full start-to-finish delivery. Depending on budget, it is often a wise investment to engage these companies.

1.7.3 Contactor Sockets

Contactor sockets, or contactors, are the interface between the DUT and load board and are often the most critical element of the production-test solution. The contactor is relatively small in size (compared to the rest of the hardware), but infinitely large in value. There have been numerous incidences where more than a million dollars' worth of production ATE and handler equipment have been interfaced with an expensive load board only to have a poorly designed contactor enfeeble the entire setup. Compounding this issue is that the redesign of a contactor can require months, which can eliminate any possibility of ever meeting the device time-to-market window.

There are various types of contactor technologies, corresponding to the style of package to be tested. They are mechanical and exercised with each DUT that is placed onto the load board, and they have a limited lifetime. Contactors are usually a removable assembly that is mounted onto the load board. When selecting a contactor, make sure that the contactor can be replaced quickly and easily, as it will be replaced frequently on the production-test floor.

When choosing a contactor it is essential to meet certain electrical, mechanical, and temperature performance requirements. From an electrical perspective, the contactor must be able to withstand high power and provide minimal distortion to high-frequency signals. In the case of testing RF power amplifiers, where high currents may be used, special contactor materials and large heat sinks may be used. This means that they introduce low inductive and capacitive impedances and provide a low contact resistance. They must also be mechanically reliable to be able to withstand many insertions. Consider that a test that is executed in one second could contribute to more than 80,000 insertions per day. Currently, typical contactor lifetimes are on the order of 1 to 2 million insertions (that could be less than 1 month). Additionally, if the DUT is to be tested at various temperatures, contactors must provide thermal insulation to maintain the DUT at a constant temperature and be able to change temperature without developing condensation that could affect the measured values of a test.

There are cost-accuracy trade-offs with contactors also. If utmost accuracy of measurements is needed, it may be necessary to select an expensive contactor with a low lifetime (low number of insertions). On the other hand, if accuracy is not the most important parameter and maximum throughput is, then a lower-cost contactor with a long lifetime may satisfy the requirements. Regardless, with any combination of the above, all of the costs of the contactor, replacement downtime, and frequency of replacement must be considered.

Particularly with discrete RF devices, but also with RF or high-frequency inputs to an SOC device, it is important to have the physical size of the contactor be as small as possible. This is because it will allow the placement of impedance matching inductors and capacitors close to the DUT. In a few cases, manufacturers produce oversized contactor housings, but they have material removed from the underside so that matching components may be placed close to the DUT.

For engineering and characterization purposes it is often desired to have a contactor with a clamp, or hold-down, on it so that a test engineer may manually place a DUT onto the load board. This is critical during load board debugging as impedance matching can be performed on the load board without having to work around the handler.

1.7.4 Production RF and SOC Wafer Probing

Another method of interfacing to the DUT is via wafer-probing equipment. Wafer probing ensures that the chip manufacturer avoids incurring the significant expense of assembling and packaging chips that do not meet specification by identifying flaws early in the manufacturing process. Small radio frequency integrated circuit (RFIC) devices in low-cost packages have traditionally been packaged with little or no RF testing (often times without a dc functional test) [1]. RF testing was done only at final test, since package scrap costs are very low. As integrated circuit (IC) complexities increase, yields become lower and the package costs higher, creating a need for screening before packaging to minimize wasting packages. As integration levels continue to rise and package complexities increase, package inductance requirements demand chip-scale packages (CSPs) or flip-chip assemblies. This requires the delivery of what are referred to as known-good-die (KGD). Furthermore, many of these RF and SOC ships are packaged in expensive multichip modules (MCMs), requiring KGD screening in production at microwave frequencies. In this case, bare dies are sold to an integrator. The integrator purchases different die types from different vendors and then integrates them all into one package. Wafer probing is mandatory in situations like this. In the early 1990s, production microwave and high-speed ICs for expensive modules or packages were being fully RF probed before assembly. In the late 1990s, consumer devices for wireless communications began to be wafer probed routinely [2].

Surprisingly, even though there are still many difficulties, many RF tests can be performed with wafer probing. Reference [3] provides extensive detail on performing many of these measurements. Table 1.1 lists just some of the measurements that can be performed with wafer probing.

Production RF wafer probing differs from traditional bench top wafer probing in that a probe card is required. A probe card, serving the purpose of the load board and contactor (in an analogy to package testing), is a complex printed circuit board that contains a customized arrangement of probe needles or probe tips to allow all of the necessary tester resources to contact all of the bond pads on one or more chips simultaneously. While there are many types of probe cards available for production testing, only a few are suitable for use at microwave frequencies for wireless communications die testing.

The performance of the probe card is sometimes the least understood section of the entire measurement system. Much effort should be expended in controlling parasitics and bypassing and controlling impedances in designing a probe card so that it works to its maximum performance. RF probe card options are limited to blade needle cards with coaxial probe blades or membrane-style probes [4]. Coaxial blade cards are able to contact three or four widely spaced single-ended RF ports through 110 GHz, but have poor ground and power bypassing parasitics. Above about 1 GHz, membrane-style probes are the only option offering high density, low power, and ground impedances or element integration close to the IC pads. Needle and coaxial probe cards do not allow bypass capacitors to be placed close enough to the DUT, and when placed closely, there is still a considerable amount of lead inductance between the device and the bypass capacitor. The membrane probe allows low impedance microstrip lines to connect bypass capacitors between power and

Table 1.1 RF Measurements Performed with Wafer Probing

Adjacent channel power	Sensitivity
Complex demodulation	Mixer conversion gain or loss
Digital input-threshold voltage	Mixer leakage
Digital output levels	Noise figure
Power-added efficiency	Intermodulation products
Frequency accuracy	Phase noise
Frequency versus time	Power pulsed power
Gain	S-parameters
Gain compression	Spurious signals
Harmonic distortion	Switching speed
Digital modulation quality	VCO frequency
Isolation	VSWR

ground. The ground inductance on the membrane card is sometimes an order of magnitude less than other types of probe cards.

Interfacing with the probe card assembly to the tester can be accomplished by any of the following:

- Cabling from the tester to the probe card;
- Use of a probe interface board (PIB);
- Direct mating of probe card to tester.

Cabling from the tester to the probe card is the least mechanically complex method of interfacing. This is also the lowest-cost solution because it allows the flexibility of probing different types of DUTs without the need for DUT-specific mechanical, or docking, hardware. If the tester has a large test head that requires a manipulator, this technique does not require alignment and mating to the wafer prober, which saves time between lot changes. From an RF measurement standpoint, there is usually some cable loss associated with this type of interface, as well as a risk of having intermittent connections at the connectors if they fail or are not tightened properly.

A prober interference board (PIB) is a mechanical fixture that ties the tester load board and the probe card together. The biggest advantage of this technique is the amount of load board space that becomes available. Any custom circuitry that is critical to being close to the test head, but not to the DUT, can be placed on the tester load board, which can be application or device specific. A pogo-pin assembly typically accomplishes dc and low-frequency ac connections to the test head. While PIB setups are the most flexible, they are also the most costly solutions, as a load board, mechanical docking hardware, and the probe card are required. The initial cost is often outweighed by the reliability and the segmented assembly that allows sections to be interchanged when repair or replacement becomes necessary.

Of the three techniques, direct mating of the probe card to the tester has the inherent advantage of being able to provide the lowest loss because of the direct connection to both the tester and the wafer prober. The test cell setup of this technique is both efficient and reliable, thus, making it a good choice. However, the disadvantage of this type of interface is that the only place for supporting components for the IC or circuitry needed to customize the test is on the probe card. In addition, there is also no mechanical isolation between the test head and the probe station.

The wafer probing station, or wafer prober, is the robotically controlled equipment that handles the wafers. There are only a handful of manufacturers, which produce these for production use. The probe stations are also available with automated wafer handling, calibration functions, testing devices at temperature, low-noise environments, automatic probe-to-pad alignment, and software integration to the tester.

If wafer probing is to be performed in production testing, some foresight and planning must occur. Additional contact points may have to be designed onto the chip for the probe to land on at test time. This can add to the cost considerations.

Making the choices involved in performing production RF wafer probing can be initially overwhelming. Many ATE vendors offer full solutions or consultation services between the wafer probe equipment manufacturer and the end user to simplify the tasks involved.

1.8 Calibration

Whether the production tester is of the rack-and-stack or ATE type and whether the interface is via handler or wafer prober, there is a need to ensure that the obtained measured values are based on calibrated measurements. There are multiple stages and purposes of calibration.

First of all, as it will be shown in Chapter 4, the power measurement is the most fundamental and serves as the basis of RF and SOC measurements. Therefore, it is imperative to have a common calibration basis point. The calibration must be traceable to something that is recognized by a general international audience so that valid comparisons can be made. For most production testers of any type, that basis is the National Institute of Standards and Technology (NIST). NIST is the generally recognized body that creates these traceable standards.

Numerous papers and books have been written describing the multitude of methods used to calibrate for RF power measurements. Other than a brief overview of on-wafer calibrations, no attempts to enhance that body of work will be presented in this book. Instead, a few general references will be offered to the reader [5–7]. The remainder of this book will assume that a NIST-traceable calibration has been performed before proceeding with any of the measurements. The NIST-traceable calibration may be made simply to the load board or all the way to the contactor socket.

Another (but not always necessary) type of calibration is termed *de-embedding*. Although used mostly for wafer probing, it can also be performed for packaged-part testing. De-embedding calibration requires the use of additional standards that are replicas of the device (wafer probing) or package (package testing). There are at least four standards (at a minimum): short, open, 50-ohm load, and through connections. With RF probing, it becomes necessary to perform this additional calibration to compensate for every component all the way to the probe tip. These standards can be readily produced though a combination of the device designer's knowledge of the device and the help of probe card models supplied by the probe card manufacturer. In contrast, for packaged devices, special standards must be designed and fabricated in the package type that is used for the device. This is a custom and expensive operation that is not highly utilized for a final production solution as it adds another process step, which increases the already high cost of test. In addition, it is often error prone whereas generally most of the errors can be accounted for during the correlation stage. Most ATE testers provide the ability to perform de-embedding calibration of both die and packaged parts.

Finally, ATE and rack-and-stack testers should be subject to an overall calibration. This is usually performed with a frequency that is based upon the ATE's manufacturing process and experience. Also, whenever periodic maintenance or replacement of any tester hardware occurs, it should be followed with a calibration. With RF frequencies, the mistake of forgetting to torque a connection properly can make a difference in accurately assessing a DUT.

1.9 The Test Floor and Test Cell

The test floor is where all of the production testing takes place. The test floor is usually a clean-room environment, free of dirt as well as electrical noise, and where as much electrostatic discharge (ESD) precautionary measures as possible are taken. One anecdotal comment from our experience is that it is very surprising to see the large number of test floor environments that take all of the precautionary measurements, but neglect to ban the use of mobile phones in the area of testing. Emissions from mobile phones create interference that can either lead to passing a bad part or failing a good part. Neither case, of course, is desirable.

The test cell is the area surrounding a test system. At a minimum, an ideal test cell consists of the test system, a handler or prober, an ESD-safe table for organizing tested and untested lots of devices, a hardwired telephone (not mobile or cordless), and provisions for air and vacuum (for running the handler or wafer prober). Additionally, if low-noise measurements are being performed, an electronically shielding enclosure on the load board, or a screen room, may be needed.

1.10 Test Houses

With the increasing trend of outsourcing processes outside of a company's core competency, the outsourcing of production testing of RF and SOC devices is also gaining popularity. Furthermore, the avoidance of the risk associated with purchasing expensive capital and the possibility of having it sit idle during market fluctuations makes this concept even more attractive. Using a pay-per-use philosophy, semiconductor manufacturers can use what is termed a test house. This is a facility that is fully equipped to provide its customers with the equipment and resources for their testing needs. The added benefit arises from the test houses having their own personnel for operating and maintaining the test systems. These costs are absorbed into their hourly rates of usage (see Chapter 3).

It is important to note that there are some liabilities associated with test houses (outsourcing). In order to monitor contracts effectively, a company should still have at least some expertise in-house that can establish contract specifications that make sense, ask appropriate questions, monitor progress, and work as a partner with the test house to overcome problems.

1.11 Accuracy, Repeatability, and Correlation

There are three critical concepts that need to be addressed when setting up a production-test plan. They are accuracy, repeatability, and correlation. In various areas of production testing, each concept has its respective importance, and often one has to be traded off for the others.

Accuracy, which pertains to production testing, is how well the results of a test are in agreement with the actual value. Accuracy is critical when a specific piece of information is needed from a test. However, incredulously, accuracy may not always be the most important target of a production test.

Repeatability is often far more important than accuracy. For example, if a very low-powered signal is to be measured and a repeatable result is found, although it is slightly off from the expected value, this may be deemed acceptable. If the results are logical, or near the expected value, then, provided a constant offset can be determined, this is also acceptable. Ambient noise due either to the test system or the environment is often the cause of repeatability problems [8]. In addition, mechanical wearing of connectors, contactor sockets, or wafer probe contacts can lead to repeatability problems.

When an acceptably accurate and repeatable value has been obtained from a production test, it is then critical to ensure that the results are not fine-tuned to the specific test system. There must be some correlation to the bench top (laboratory) measurements. And furthermore, if a test house is used, the results must agree between the many different test systems that the test will be performed on. This is essential for minimizing the introduction of errors into the production tests.

1.12 Design for Testing

The commonly used acronym, DFT, with reference to test engineering, stands for design for testing (this usage of DFT should not be confused with the mathematical algorithm, the discrete Fourier transform, which may also be used in some test plans). DFT refers to the scenario where the design engineer of the device has an understanding of production testing and is aware of the specific needs of the specific device. With this information, the design engineer can make provisions to facilitate production testing and lower the overall cost of test. An example would include creating an external package pin that is never used in the product's final application, but only during testing. A common pin name used for exactly this case is "TEST."

Common discrete RF devices for wireless communications are two- and three-port devices. For RF devices, all of the necessary access is available. All of the RF test parameters can be fully determined by applying signals at the DUT's ports. Hence, DFT has not been a large topic when discussing production testing of purely RF devices.

DFT is utilized more often in SOC devices where, for example, the intermediate RF stages of an RF-to-baseband SOC device may need to be accessed for testing. In the intended use of this particular SOC DUT, however, the pin may serve no purpose. A simple example of DFT is the received signal strength indicator (RSSI) pin[2] on an SOC receiver. RSSI provides a dc voltage signal that is proportional to the strength of the RF power being received. This signal is very helpful, because the package of an SOC receiver does not allow access to the RF signal.

Because of the high levels of integration of SOC devices, it is inevitable, as production testing becomes an increasing percentage of the overall device cost, that the trend to add features (DFT) will grow.

1.13 Built-in Self-Test

Built-in self-test (BIST) is very common in highly complex digital devices and memory devices, where at the device design level circuits are built into the device that are

---

2. Although the RSSI signal is integral to the function of the device, the external access to it via a pin may be considered DFT.

only used during testing. They often serve no other functional purpose in the end application. Innovative attempts are being made to pioneer integrated test circuits into the design of analog devices, even RF devices. BIST could reduce the quantity of tests that are needed. For example, currently, in an SOC transceiver, digital signals of the device are monitored and analyzed to determine whether the device is in the transmitting or receiving state. BIST designed into the device could potentially indicate information and eliminate the need for tests such as turn-on time or lock time.

References

[1] Strid, E., "Roadmapping RFIC Test," *1998 GaAs IC Symposium Technical Digest*, pp. 3–6.

[2] Gahagan, D., "RF (Gigahertz) ATE Production Testing On-Wafer: Options and Tradeoffs," *Proceedings of 1999 International Test Conference*, p. 388.

[3] Wartenberg, S., *RF Measurements of Die and Packages*. Norwood, MA: Artech House, 2002.

[4] Lau, W., "Measurement Challenges for On-Wafer RF-SOC Test," *Proceedings of International Electronics Manufacturing Technology Symposium*, 2002, IEEE Cat. No. 02CH37299, pp. 353–359.

[5] Wong, K., and Grewal, R., "Microwave Electronic Calibration: Transferring Standards Lab Accuracy to the Production Floor," *Microwave Journal*, Vol. 37, No. 9, 1994, pp. 94–105.

[6] Dunsmore, J., "Techniques Optimize Calibration of PCB Fixtures and Probes," *Microwaves & RF* Vol. 34, No. 11, 1995, pp. 93–98.

[7] Fitzpatrick, J., "Error Models for Systems Measurement," *Microwave Journal*, Vol. 22, No. 5, 1978, pp. 63–66.

[8] Burns, M. and Roberts, G., *An Introduction to Mixed-Signal IC Test and Measurement*. Oxford: Oxford University Press, 2001, p. 18.

CHAPTER 2
RF and SOC Devices

2.1 Introduction

A few years ago, an RF applications/test engineer would have been very focused in this very unique (often termed complex) field, performing tests on discrete RF devices such as mixers, power amplifiers, low noise amplifiers, and RF switches. Today, the test industry faces increasing levels of integration such that many of these discrete device functions are used as building blocks and are contained within one chip or module. Furthermore, the integration levels are such that system-on-a-chip (SOC) devices contain baseband (analog) functionality as well as digital functionality (earlier RF devices often contained three-wire serial communications for controlling things such as gain control, but current digital functionality of these devices is becoming more complex).

Figure 2.1 shows a typical wireless digital radio, which is the foundation for many consumer devices such as mobile phones, cordless phones, pagers, and wireless LAN (WLAN) radios. It is apparent that many components are needed to bring signals into and out of the underlying microprocessor that acts as the brain of the device. While many components make up a complete wireless digital or analog radio as used in today's telecommunications industry, this book will focus on the following:

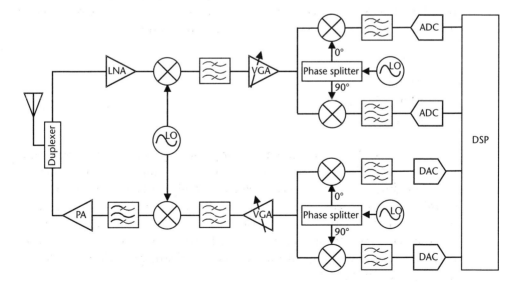

Figure 2.1 Typical wireless radio block diagram.

- RF low noise amplifier (LNA);
- RF power amplifier (PA);
- RF mixer;
- RF switch;
- Variable gain amplifier (VGA);
- Baseband/IF modulator;
- Baseband/IF demodulator;
- Transmitter;
- Receiver;
- Transceiver.

SOC devices are those that have more than one of the above devices combined on a substrate to provide some function, for example, placing all of the devices that make up a mobile phone handset onto a single microchip. Over the past few years, there have been many attempts to place the complete wireless radio on a chip, but for practical reasons, what is termed SOC is often only a portion, such as that comprising the input/output at the antenna down to the analog baseband input/output on a wireless transceiver. Thus, an exception to the above statement is that a discrete transmitter, receiver, or transceiver may also be termed an SOC device. The recent trends have been moving toward much higher levels of integration. This is primarily due to two reasons: reduced-cost at the consumer level and the desire for reduced power consumption (longer battery life). It is apparent that lower-frequency analog and lower-level digital functionality is coresiding on the SOC chip with RF front-end devices. This trend will continue as pressures to achieve the above two goals surmount.

SOC devices, as used in this discussion, have at least one RF input (or output). Based on that, SOC devices for wireless communications can be broken down into the following types, based on input/output configuration:

- RF/RF;
- RF/IF;
- RF/baseband;
- RF/digital.

RF/RF and RF/IF are treated similarly with respect to testing procedures. The measurement techniques for IF frequencies still require attention to detail and an understanding of making measurements at high frequencies where traditional Ohm's Law–based calculations will not work. Examples of these types of SOC devices would include a chip consisting of a filter/LNA combination or filter/LNA/mixer combination to be used as the front end of a receiver. Additionally, they may have some digital signals for received signal strength indicator (RSSI) or automatic gain control (AGC).

RF/baseband SOC devices are used quite commonly today in WLAN modems. The may contain everything (for a receiver, for example) from the input filter/LNA all the way to the in-phase, quadrature-phase (I/Q) outputs. When testing these devices, the engineer must have an understanding of RF measurement techniques, which are based on the frequency domain, and also have an understanding of making measurements in the time domain.

RF/digital SOC devices are used quite commonly today in Bluetooth modems. The reason for this is that the Bluetooth architecture is relatively simple to implement on a single chip. It has been explored quite exhaustively, and as a result the low cost pushes a minimum number of chips to be used in a Bluetooth modem.

Baseband/digital devices also fall under the category of SOC; however, from a testing point of view, these devices fall under the category of full mixed-signal devices. There are numerous resources available on the topic of testing mixed-signal devices, such as [1].

This chapter is intended to introduce the reader to the various types of discrete RF and SOC devices. The following sections of this chapter will provide an overview of each of the SOC components and then bring together the full SOC receiver. Examples are based upon the superheterodyne receiver, but they apply equally to the zero-intermediate frequency (ZIF) receiver. Afterwards, a brief overview of each of the two-transceiver architectures will be presented and a comprehensive listing of tests that are performed on each of the RF and SOC devices will be provided.

2.2 RF Low Noise Amplifier

The low noise amplifier (LNA) is often the most critical device in the receiver chain of a wireless device. The LNA must amplify the extremely weak signals received by the antenna with large amounts of gain, while minimizing the amount of added noise. Since it is the first device that processes the incoming signal, it is critical that its additive noise be extremely low [see Friis equation, (8.13)]. Thus, the noise figure (NF) of the LNA is often the most difficult and critical parameter that is tested in production. From a design point of view, the difficult task is to provide high gain while minimizing the introduction of noise. These two items are historically mutually exclusive. Figure 2.2(a) shows the block diagram representation of an RF low noise amplifier.

2.3 RF Power Amplifier

A discrete RF power amplifier (PA) is required at the output of a transmitter and is the one discrete device that often continues to remain a stand-alone discrete device (although for many Bluetooth and low-power wireless networking device

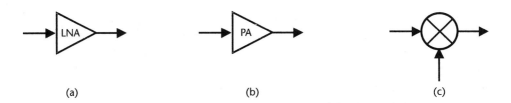

Figure 2.2 Block diagram representations of RF devices: (a) low-noise amplifier, (b) power amplifier, and (c) mixer.

architectures, the PA is integrated). PAs are used at the output of a transmitter to boost the signal level so that it can reach its final destination, which may be a great distance away. There are many reasons why PAs still remain largely discrete devices and have not been integrated into SOC devices: PAs must be able to "boost" the transmitted signal to a relatively high power for it to traverse the long distance and be successfully received. This requires a rugged amplifier with lots of gain. Additionally, the PA is often pulsed (for example, GSM), which requires that the PA must also have fast response times. The large gain and fast response time requirements invariably mean that PAs produce large amount of power and generate large amounts of heat. Furthermore, PAs are normally in the 20% to 30% efficiency range and, thus, they drain the battery considerably. All of these requirements dictate that a specialized manufacturing process for PAs be used. This specialized manufacturing process is very different from that of other RF devices, often requiring hybrid semiconductors such as gallium arsenide (GaAs) and silicon germanium (SiGe) technologies. Chip manufacturers are constantly seeking an equivalent SiGe power amplifier that would allow integrating the PA with the rest of the SOC. Numerous efforts exist from the SiGe design community to make this combination successful [2]. Both scenarios, integrated and discrete, will likely be partially successful for different kinds of radios. Figure 2.2(b) shows the block diagram representation of an RF power amplifier.

2.4 RF Mixer

A mixer is often referred to as a frequency-translating device because its purpose is to perform either upconversion or downconversion of a signal. Acting as an upconverter, a mixer can be found in the transmit chain of a wireless or SOC device. Mixers are also used as downconverters, such as where they convert RF to IF in a receiver.

The mixer differs from the aforementioned devices with the first big difference being that a mixer is a three-port device. It has two input ports and one output port [see Figure 2.2(c) for a mixer's symbolic representation]. The three ports are usually denoted as radio frequency (RF), intermediate frequency (IF), and local oscillator (LO). The mixer is a frequency-translating device; that is, the input and output frequencies differ from each other. The fundamental operation of a mixer is based upon its intentional nonlinear products, much like the nonlinear intermodulation products of an amplifier (albeit, those are unwanted in that case). The purpose of a mixer is to "move" the incoming frequency to some other outgoing frequency or, more concisely stated, to translate f_{in} (the input frequency) to f_{out} (the output frequency). The LO port is always an input port and is used as a kind of "pump" to translate f_{in} to f_{out}. The RF and IF ports are bidirectional ports. Since a mixer has three ports, this means that it has nine S-parameters. Typically only five of these are tested in practice. They are shown in Figure 2.3. S-parameters are discussed in detail in Chapter 4. As you may expect, the mixer is one of the most critical RF building blocks because it is always operating in the nonlinear region. As such, it is difficult to design and manufacture a mixer because during the normal operation of a mixer, there are many linear frequency translations and other unwanted nonlinear frequency translations that are occurring. These other frequency translations are, of

2.4 RF Mixer

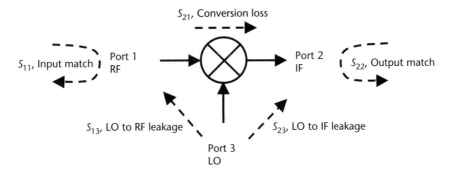

Figure 2.3 Mixer parameters and their equivalent S-parameters.

course, undesired and must be minimized and filtered. Overcoming these problems has been one of the major hurdles to the successful development of the zero-IF radio.

A mixer is made up of one or more nonlinear devices (i.e., diodes, FET transistors, and so forth) acting in their nonlinear ranges. The simplest construction of an RF mixer is the single-ended mixer as shown in Figure 2.4, along with its block diagram representation. The input RF and LO signals are combined and passed into a diode. Afterwards, a filter may be used to remove unwanted frequencies resulting from the nonlinearity of the diode.

There are several types of mixers, and each has its own purpose. Most of the more complex mixers are based upon the single-ended mixer. Table 2.1 shows the various types of mixers and their typical characteristics. Properties such as voltage standing wave ratio (VSWR), isolation, and conversion loss are described in Chapter 4.

Another common type of mixer is the double-balanced mixer. This is shown in Figure 2.5. The four diodes in a configuration similar to a bridge rectifier produce an output signal that consists only of the sum and difference frequency components of the two inputs. Because of this, a double-balanced mixer has excellent isolation (typically 50 dB at wireless-communications-device operating frequencies), meaning that neither of the two input signals appears as a component of the output signal. This is often a problem of the single-ended mixer. The power consumption is low (low conversion loss) and most designs are broadband to cover wide frequency

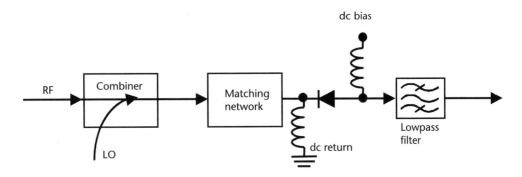

Figure 2.4 Single-ended mixer as a downconverter.

Table 2.1 Mixer Characteristics

Type	Number of Diodes	VSWR	Isolation	Conversion Loss
Single ended	1	Poor	Fair	Good
Balanced (90)	2	Good	Poor	Good
Balanced (180)	2	Fair	Excellent	Good
Double balanced	4	Poor	Excellent	Excellent
Image rejection	8	Good	Good	Good

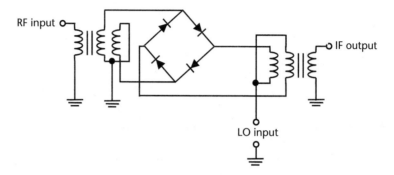

Figure 2.5 Double-balanced mixer as a downconverter.

ranges. The drawbacks of these mixers are that impedance matching at the ports is critical, so if it is being used for broadband applications, there may be difficulty in matching to maintain a constant impedance across all frequencies. Additionally, they require relatively high-powered local oscillator drive signals.

Image-rejection mixers provide an output signal that consists of the desired output at the new frequency and two image signals that are 180° out of phase of each other. Because of the 180° phase shift, they cancel. Figure 2.6 shows that the primary phase cancellation comes from the use of two 90° hybrid couplers. A hybrid coupler, more commonly just called a hybrid, is a four-port device that divides power from each of ports 1 and 2 equally among ports 3 and 4. The signals at ports 3 and 4 have a 90° phase shift between them. Additionally, no energy is transferred, or coupled, between ports 1 and 2. Each of the hybrid's output signals is then passed on to a separate path where it is downconverted (recall these signals are 90° out of phase with respect to each other) and then passed through another 90° hybrid. As a result, the two outputs of the final hybrid are referred to as the upper and lower sideband signals, absent of image signals, as the image signals end up with a total of 180°

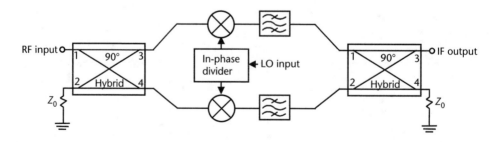

Figure 2.6 Image-rejection mixer as a downconverter.

2.5 RF Switch

RF switches are used in nearly every RF and wireless application. They are used inside phones and other wireless communications devices for duplexing and switching between frequency bands and modes.

They are typically bidirectional. RF switches come in two primary varieties, absorptive switches and reflective switches. In reflective switches, the impedance of the "off" port is *not* 50 ohms, and often a mismatch occurs, hence the name reflective. As a result, this type of switch has a very high voltage standing wave ratio (VSWR). An absorptive switch has very good VSWR in *both* the on and off modes of the switch.

The two major classes of technologies used to implement switches are *p*-type silicon/insulator/*n*-type silicon (PIN) diode switches and GaAs field effect transistor (FET)–based switches. GaAs switches can also be PIN types, but are more commonly FET-based.

Diode-based switches make use of a PIN diode. Figure 2.7 shows how a PIN diode can be used to create an RF switch. Assuming that the RF signal is small relative to the dc bias established across the diode, the diode can either be forward biased (allowing the diode to conduct with low impedance) or reverse biased (making the diode appear as an open circuit). If the RF signal becomes relatively large, solid-state switches add distortion due to the nonlinearities of the diode I-V curve. There is an upper frequency limit for PIN switches due to the parasitic junction capacitance that shunts the diode. This capacitance reduces the overall impedance seen by the RF signal in both the on and off states. If that capacitance is too large, the diode will not turn off effectively. PIN switches are often used in pulsed RF applications, as they are able to handle the high power usually required of pulsed RF signals. Typical on/off switching times of PIN diode switches are on the order of microseconds.

GaAs switches use gallium arsenide technology to create a FET, or field effect transistor, used in the nonlinear (switching) mode. The switch is either fully on or fully off, depending upon bias conditions. Switching times of GaAs FET switches

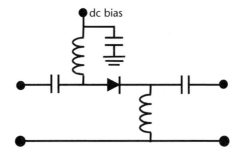

Figure 2.7 Implementations of a PIN switch.

are on the order of nanoseconds. Additionally, they have a good frequency response all the way down to dc.

A primary difference between PIN and FET switches is that PIN switches require significant dc current in the on state, while FETs consume only leakage current in both on and off states. Current drain can be a critical specification for RF switch selection and testing.

It should be noted that in addition to being a DUT, RF switches are often used in automated test equipment and on production load boards to perform switching operations when routing signals. As an example, highly integrated wireless SOC devices with multiband radios are good candidates to implement RF switches on the production-test application. Only one of the multiple radios is "on" (i.e., transmitting or receiving) at any given time, so from a strictly hardware cost perspective, it is more cost effective to employ switches than to use dedicated hardware.

2.6 Variable Gain Amplifier

It should be pointed out that wireless communications RF and SOC devices have enormous dynamic ranges. This trend of wider dynamic ranges is increasing. The further away from each other that two wireless devices are, whether they are Bluetooth devices, WLAN devices, cell phones (mobile phones) and base stations, pagers and base stations, satellite links, or any other wireless devices, the higher their output powers must be in order to sustain the wireless link between them. Conversely, if the two wireless devices are very close to one another, then their output powers must be lower so as not to overdrive or compress their wireless counterparts. The more wireless devices that are added to a specific area, (a downtown city district for example), the more confusing it becomes due to the large number of combinations of high powers, low powers, rejections, and compressions. Each wireless device must be able to change its transmitted and received power levels quickly to acclimate to its continuously changing surroundings.

This brings us to the subject of automatic gain control (AGC).[1] To simplify the discussion and reduce the number of variables, the subject of AGC will be constricted to discussing only the transmitter chain of the wireless SOC device [although AGC/variable gain amplifer (VGA) amplifiers are also used in the receive chains in wireless communications]. It should now be very apparent why a wireless device would need to change its transmitted output power level quickly and dynamically. The most common way of doing this is to design the wireless SOC device to have multiple amplifier output stages with one or more of the stages designed to have variable gain control. The gain of the amplifier (and ultimately the output power) can then be controlled by adjusting the variable gain control. The variable gain control is usually in the form of a voltage or a current. For this discussion, it will be assumed that the gain of the amplifier is voltage controlled. That is, by adjusting the particular voltage up or down to that amplifier, the gain is also adjusted up or down respectively. One question might be, How does the wireless device know what to set

---

1. The topic of AGC is associated with VGA. Whether automatic or not, a VGA has some means to adjust its power output.

the gain to? The answer is, the same way that you know or learn to speak louder or softer when speaking on a normal wire line phone. The other person tells you to talk louder if he cannot hear you. The same thing happens with two wireless devices. If a mobile phone call is being made and the mobile is very far away from the nearest base station, then the base station sends a signal to the mobile telling the mobile to "talk louder," or increase your gain. The base station may only have to send the request once if the mobile is stationary, but the base station may have to tell the mobile constantly to increase or decrease its power if the mobile is actually moving (in a car, for example).

A simple block diagram of an AGC connected to a test setup is shown in Figure 2.8. The input to the AGC is shown to be either an RF source or arbitrary waveform generator. A high-frequency input requirement to the device would dictate an RF source as the input. A low- or medium-frequency input requirement, perhaps with high impedance and with differential inputs, might require an arbitrary waveform generator as the input signal. The same holds true for the output. If the output of the device is high frequency in nature, then downconverting the output signal to an IF signal is probably mandatory before digitizing it. However, if the output signal already falls within the bandwidth and sample frequency limitations of the digitizer of the tester, then the output signal can be directly digitized. Additionally, digital control of the device is usually mandatory, so the test system must have this capability, and it must be synchronized to the measurement equipment.

It is helpful to pause here for a moment and consider the simple block diagram of Figure 2.8. The block diagram really represents the merging of the mixed-signal world with the RF world, and it is this merging and understanding that are required to test wireless SOC devices. If the device were merely RF in nature, then the input would have been connected to an RF source and the output to an RF receiver. If the device were only mixed signal in nature, then the input would have been connected to an arbitrary waveform generator and the output to a digitizer. Because the device is an SOC device, it may be connected to either or both, so the test equipment must be able to accommodate either or both simultaneously. Additionally, the application/test/product engineers must adapt themselves to working in both worlds simultaneously.

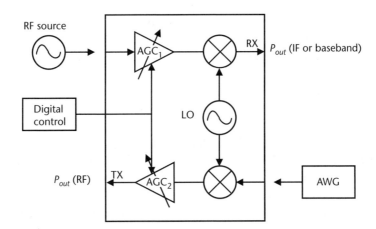

Figure 2.8 Variable gain amplifiers in an SOC device.

Wireless SOC devices have a range of gain states. It depends on the design and application of the particular device, but suffice it to say that the gain range is usually controlled by 8 to 16 bits. Even if a device only uses 8 bits of gain range, that is still 256 different gain settings. Most engineers are familiar with how analog-to-digital converters (ADCs) and digital-to-analog converters (DACs) work, but wireless amplifiers are different. Conceptually, they have a similar function, but RF amplifiers are highly dependent on their input and output impedance matching. The degree of impedance matching, in turn, is greatly dependent on frequency. What does that mean? If the AGC is programmed to a different gain level, then a different path through the amplifier may need to be used to achieve the new gain setting. A different path may have a different impedance match at that particular frequency. This means that more testing of the wireless device is suddenly required to ensure that it operates to its specifications.

2.7 Modulator

The building blocks of today's SOC devices are beginning to consist of multiples of the discrete RF devices that have already been presented. As an example, consider a modulator. A modulator can be built from discrete components consisting of two mixers, a few amplifiers, and various filtering components. However, a modulator can also be fabricated on a wafer as a single building block, and indeed, this is commonplace of today's chip manufacturers. The modulator, or I/Q modulator, shown in Figure 2.9(a), is made up of two mixers and a phase splitter. The function of a modulator is to take an incoming baseband data signal and provide an output IF signal that has the information encoded, or modulated, onto it.

The phase splitter takes an input signal from an LO and creates a 90° phase difference between the two outputs of the splitter ($\sin x$ and $\cos x$, for example). The two offset signals are then passed on to the two mixers to be mixed with the baseband signal. The outputs of the mixers are IF frequencies (in a superheterodyne architecture). The two IF signals are then combined and passed on to the front end of the system.

In a perfect world the splitter creates exactly 90° between the two signals that it outputs. For the modulation and demodulation process to function properly, it is important for the two signals to be orthogonal to one another (thus, the 90°). The receiver's bit error rate (BER) or error vector magnitude (EVM) will increase with the decreasing phase balance of the two signals. As such, production testing is often done to determine the amount of phase difference that is present. Additionally, the mixers upconvert with no disturbance to phase or amplitude, thus proving that two equal-phase and equal-amplitude signals are combined at the IF frequency.

When phase and amplitude distortion are introduced by the modulator in such a fashion, the carrier (LO frequency), which would ideally be canceled upon mixing, now appears in the IF signal. Carrier and sideband suppression testing are performed to determine the impact of this distortion. These are the two most important measurements that can be made on a modulator. Specifically, the carrier rejection indicates the amount of the unwanted carrier present at the output, and the sideband rejection indicates the phase and amplitude balance of the mixers.

2.8 Demodulator

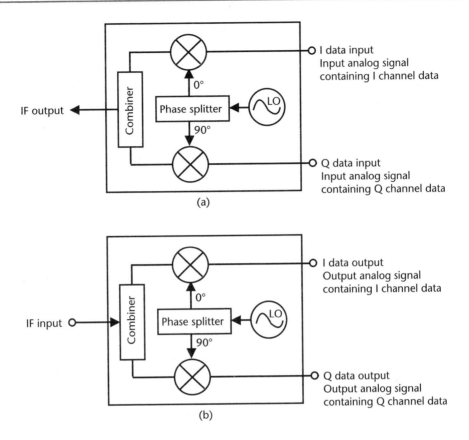

Figure 2.9 Block diagram of (a) modulator, and (b) demodulator.

2.8 Demodulator

The demodulator is analogous to the modulator and can be thought of as a modulator working in reverse. A demodulator takes an IF or RF signal as the input and produces two baseband signals denoted as I and Q, which are shown in Figure 2.9(b). The I indicates in-phase and the Q indicates quadrature-phase. Often the modulator and demodulator functions can be had from the same device.

As with the modulator, it is critical that minimal distortion is introduced in the form of amplitude and phase imbalance of the signal. Since the products of the demodulator are the I and Q baseband signals, the critical tests that are performed are I and Q amplitude and I and Q phase-balance measurements. Both, modulators and demodulators, have two basic designs, single ended or differential ended. Single-ended demodulators have only two outputs (I and Q). Differential-ended demodulators have four outputs (I, $\overline{\text{I}}$, Q, and $\overline{\text{Q}}$). Differential modulators/demodulators are growing in popularity due to their superior noise-reduction properties over single-ended designs. To better understand the basic tester hardware requirements necessary to test a demodulator, consider the following example. Let the input to a demodulator is a single continuous wave tone. The output I and Q signals ideally should have equal amplitudes and are orthogonal to one another. To test the amplitude and phase imbalance of the two signals, a test system must be able to capture both signals simultaneously. This can be realized either with a single digitizer with a dual core or with two separate digitizers that are time synchronized. Most systems on the market offer at least one of these hardware configurations.

2.9 Transmitter

An SOC transmitter is pictured in Figure 2.10(a). In general, a transmitter implies that its output is the signal that is RF and ready to be sent to a PA or directly to the antenna for transmission. The function of a transmitter in a wireless communications device is to take the data-containing signal, modulate it, and then send it to the antenna. As discussed in Section 2.3, the PA is typically a separate device and continues to be due to the different materials and processing that are used.

One of the primary functions of the transmitter is that of the modulator. It is therefore not surprising that critical tests of transmitters are carrier and sideband suppression. Additionally, measuring the output power spectrum at inband and out-of-band frequencies of its intended use to locate any spurs is also critical to ensure that the device complies with the specifications to which it was designed.

2.10 Receiver

The purpose of the receiver shown in Figure 2.10(b) is to amplify the weak RF signals that are received at the antenna and downconvert them to much lower

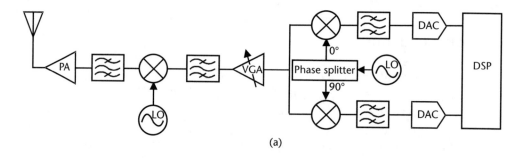

(a)

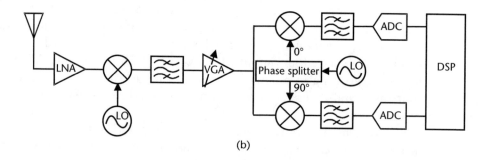

(b)

Figure 2.10 Block diagram of a superheterodyne: (a) transmitter, and (b) receiver.

frequencies that can take advantage of less costly low frequency electronics to extract and process the information. With the superheterodyne receiver, multiple stages of gain and downconversion are employed.

The incoming signal first is filtered to remove any unwanted RF energy that is outside the frequency band of interest. Then it is passed onto the LNA to amplify the signal while introducing minimum additional noise. Finally, the signal is downconverted with mixers to an IF frequency (typically less than 300 MHz) and converted to I and Q baseband signals through a demodulator. These are then to be sent to analog-to-digital converters (ADCs) that generate digital patterns to be processed by subsequent baseband/digital signal processing (DSP) functions. The DSP is often contained in a separate chip known as a baseband processor.

One of the most critical test items for SOC receivers is sensitivity. Sensitivity is an extension of the discrete RF device noise figure. Noise figure is a measure of sensitivity. The higher the noise figure, the harder it will be for a device to receive low-level signals. With SOC devices, the all-encompassing term *sensitivity* is used because it is possible to measure their full functionality. For instance, it is possible to provide modulated RF signals to the SOC device. The device then processes the signal and produces either analog baseband signals (in-phase and quadrature-phase voltages) or digital data (bits) at the output.

2.11 Transceiver

Building further upon the items that we have just discussed, the SOC transceiver pictured in Figure 2.11 consists of both transmitter and receiver functionality. The transmitter and receiver functions in the transceiver can either work simultaneously or in switched mode, whereby the device is either only transmitting or only receiving at any given time (this would, of course, require two antennae or a duplexer).

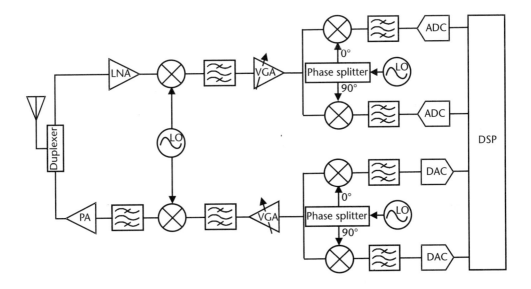

Figure 2.11 Superheterodyne transceiver.

All of the items that are tested for both the SOC transmitter and the SOC receiver are tested in the SOC transceiver, as one would expect. There are some additional tests that may arise depending upon the specific design of the transceiver.

A transceiver can have a common LO that is shared between the transmit and receive paths. When this is incorporated, it becomes necessary to test each chain for signals that may arise due to the other chain. Additionally, if any form of duplexing is used at the front end of the device, then it also becomes necessary to test for TX/RX leakage.

As with either the SOC transmitter or SOC receiver, the signals coming from the non-RF end of the device may be either digital or analog. The tests that must be performed on these are the same as those discussed in Sections 2.9 and 2.10.

2.12 Wireless Radio Architectures

The preceding sections have based their descriptions upon the superheterodyne radio architecture. Wireless radio can incorporate various designs; however, the two most common are the superheterodyne and the zero-IF (ZIF). The superheterodyne structure is more prolific, but the ZIF has recently emerged due to new technological advances in the radio design industry.

2.13 Superheterodyne Wireless Radio

In 1901, Guglielmo Marconi successfully transmitted a signal wirelessly across the Atlantic Ocean, and ever since, wireless communications has been developing. Today's wireless radios have come a long way from the initial work of Heinrich Hertz, who created a spark across a gap and received it at another gap.

The superheterodyne radio was invented by Edwin H. Armstrong in 1918 and first introduced to the market place in the mid-1920s. Because the superheterodyne radio was better performing and cheaper than the tuned radio frequency (TRF) transceiver, by the mid-1930s, it had become the de facto standard in radio design [3].

The superheterodyne transceiver is considered the classic radio architecture in which the received signal is downconverted to baseband frequency in two stages. The incoming RF signal is first downconverted to an intermediate (IF) frequency. This allows image suppression and channel selection by filtering out any unwanted signals. The filtering is commonly accomplished by use of surface acoustical wave (SAW) or ceramic filters. The filtered IF signal is then further downconverted to the baseband frequency, which is then digitized and demodulated with DSP. Because the radio has two stages of downconversion, it is generally more complex and more expensive due to the extra components like discrete SAW filters and voltage controlled oscillators (VCOs)/synthesizers. Figure 2.11 shows the superheterodyne receiver.

2.14 Zero Intermediate Frequency Wireless Radio

In contrast, the homodyne, or zero-IF (ZIF), radio transceiver is a direct-conversion architecture, meaning that it utilizes one mixer stage to convert the desired signal

2.14 Zero Intermediate Frequency Wireless Radio

directly to and from the baseband without any IF stages and without the need for external filters. A block diagram of a ZIF radio is shown in Figure 2.12, where it can be noted that there are fewer components than in the superheterodyne radio. It is also common to integrate the LNA, or low noise amplifier, VCO, and baseband filters onto one single die. ZIF transceivers are not a new concept, and they have been used for years in cellular and pager applications. They are also beginning to emerge in WLAN applications, which play an important role in the SOC market. ZIF transceivers, although cheaper, have some inherent problems that must first be overcome before employing the technology [4].

Some of the common RF problems inherent to the ZIF architecture are dc offset, flicker noise, and LO pulling. The dc offsets are mainly generated by the LO leakage, which self-mixes, thereby creating a dc component in the signal chain. In contrast, the superheterodyne architecture has filters in the IF stage that eliminate this problem; however, dc offsets affect the receiver performance and can cause the RF stage to saturate. The dc offset problem in ZIFs can be addressed by designing a compensation scheme. The compensation scheme needs to measure the dc offset and then subtract it from the signal. This method is similar to noise cancellation, where the algorithm tracks the background noise and subtracts it from the original signal, thereby improving the signal to noise ratio.

$1/f$ noise, or flicker noise, is low-frequency device noise that can corrupt signals in the receiver chain. Flicker noise is more pronounced with the ZIF architecture because of the direct conversion to low-frequency baseband. Again, in contrast, the superheterodyne architecture eliminates this problem through proper filtering. Another concern with direct conversion is the pulling of the LO by the power amplifier (PA) output, which affects the direct upconversion process. This is because the high-power PA output, which has a spectrum centered around the LO frequency, can disturb, or "pull," the VCO.

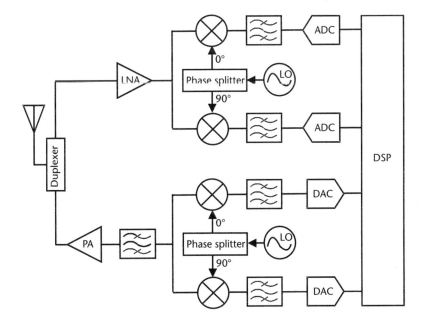

Figure 2.12 Zero-IF transceiver.

Due to recent progress in technology, these issues can largely be compensated for, which is why we are seeing more and more ZIF radio structures emerging in wireless applications, and it is likely that this trend will continue.

2.15 Phase Locked Loop

Aside from the fundamental building blocks presented earlier in this chapter that may be integrated to form an SOC device, there is one other component that is becoming more common in SOC devices. That is the phase locked loop (PLL). The PLL is a frequency-synthesis and -control circuit found on the SOC. It originated in the 1930s to work with the zero-IF receiver to allow a large number of frequencies to be synthesized from one circuit. The primary purpose of a PLL is to provide multiple, stable frequencies on a common time base within the same system. A basic PLL consists of a reference oscillator, phase detector, loop filter, and a VCO, as shown in Figure 2.13(a).

PLLs have been used commercially for many years, beginning in the 1940s when they were used to synchronize the horizontal and vertical sweep oscillators in television receivers. There, they are used for frequency-shift key (FSK) modulation and FM demodulation, but with reference to SOC devices, they are generally used for frequency synthesis and frequency multiplication and division. Their most common use in SOC devices is to generate the local oscillator (LO) frequency used with the mixers in either superheterodyne or zero-IF transmitters or receivers.

Figure 2.13(a) shows that the basic principle of operation of a PLL begins with the goal of achieving a stable frequency at the output of the VCO. In this simplest description, the VCO is first set to output a signal with a frequency equal

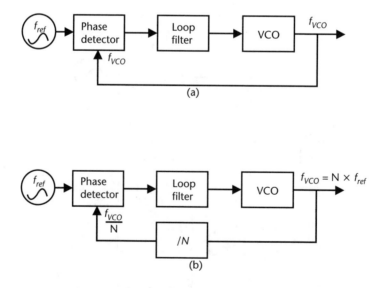

Figure 2.13 Components of a PLL circuit: (a) basic phase locked loop, and (b) fractional-N phase locked loop.

to that of the crystal oscillator reference. This is sent to another part of the SOC to be used for some function like providing an LO to a mixer. The foundation for the PLL circuit is the reference oscillator. It is usually a very stable crystal oscillator of the desired output frequency. It is used as a reference for the phase detector, which compares the phase of the reference oscillator with the negative feedback from the VCO output. If there is a difference in phase between the reference oscillator and the feedback from the VCO, the output voltage of the phase detector will cause the VCO to deviate toward the reference oscillator frequency until it "locks" onto it. A filter (termed a *loop filter*) is used to smooth the pulsed output of the phase detector so that the VCO can receive a clean control voltage. This entire cycle happens very quickly, typically within tens of microseconds. In production SOC testing this lock time, called the synthesizer lock time, is a common measurement.

That explains how to achieve a stable output signal. Now, the obvious question is, Why wouldn't one simply use the reference oscillator to achieve a desired frequency? Actually, that could be done, but it would not be very practical. If, for example, a signal were needed that could be used as an LO input to a mixer that downconverted a signal in a system having more than one channel, the system would require one PLL circuit for each channel (also one reference oscillator for each channel). This architecture would become expensive in both money and space. A solution to this is shown in Figure 2.13(b), where an integer divider is placed into the negative feedback path from the VCO. The divider divides the frequency of the VCO output by N and feeds the new signal back to the phase detector. Assume, for example, that the operating frequency of the VCO is N multiples of the crystal oscillator frequency. Then, the phase detector input from the VCO, after dividing by N, is once again the frequency of the crystal oscillator. The phase detector then stabilizes the signal to that of the crystal oscillator. Essentially, this method tricks the phase detector into believing that the VCO output frequency is at that of the crystal oscillator. This method allows the use of only one crystal oscillator, and the entire system is based upon the frequency of the same, stable crystal oscillator.

The divider is typically digitally controlled and can have any number of values. For wireless communications systems, an upper limit of about 30,000 is placed on N due to the introduction of excessive phase noise (see Chapter 8) from frequency multiplication by N [5].

PLLs have three states of operation:

1. Free running;
2. Captured;
3. Phase locked.

The free-running state describes when the output of the VCO has not yet locked onto the same time base as the reference oscillator and is not within $\pm 2\pi$ of the reference oscillator phase, meaning that they are quite different in phase and frequency. In this stage, the output of the phase detector is pulsed, with a pulse period proportional to the frequency difference. The captured state is when the VCO (or divided VCO feedback) is at the same frequency as the reference oscillator, but the phases differ. At this point, the output of the phase detector exhibits a dc offset voltage. Finally, in the phase-locked state, the VCO output signal has been adjusted to the same, stable time base as that of the reference.

2.16 RF and SOC Device Tests

Table 2.2 shows many of the common parameters that are tested in RF and SOC devices. The table is not all inclusive, as many test items on a device are dependent upon the specific manufacturer, but it may be used as a guide to provide an estimate of what will be involved in writing a test plan.

Some tests referenced in this book include BER and EVM. These types of tests are also referred to as system-level tests. The term *system level* is used because we are not discussing the testing of all of the discrete components that make up the SOC, but rather we are viewing the SOC as more of a traditional black box.

Table 2.2 Production Tests on RF and SOC Devices

Parameter	LNA	PA	Mixer	Switch	Transmitter	Receiver
VSWR		X	X	X		X
Return loss	X	X	X		X	X
Insertion loss				X		X
Gain	X	X				X
Gain flatness		X				X
Isolation	X		X	X		
Linearity		X				
Noise figure	X	X	X			X
Dynamic range		X				X
Power compression (e.g., $P_{1\,dB}$)	X	X	X	X	X	X
Third-order intermodulation product (IP3)	X	X	X		X	X
Third-order intercept point (TOI)	X					X
Harmonic distortion		X	X			X
Conversion loss/gain			X			
Intermodulation distortion			X			
Switching speed				X		
Bandwidth	X	X			X	X
Power-added efficiency (PAE)		X				
Spurious output		X	X			X
RF-LO rejection			X			X
ACPR/ACLR			X		X	
Phase noise					X	X
I/Q offset						X
I/Q amplitude match						X
I/Q phase match						X
Output power					X	
Carrier suppression					X	
Error vector magnitude (EVM)					X	X

References

[1] Burns, M., and Roberts, G. W., *An Introduction to Mixed-Signal IC Test and Measurement,* Oxford: Oxford University Press, 2001.

[2] "IBM SiGe Technology Improves Power Amps for Cell Phones," IBM press release, June 14, 2002.

[3] Poole, I., *The Superhet Radio Handbook*, London: Bernard Babani Publishing, 1994.

[4] Saidi, M., "Dual-Band Issue: Super-Heterodyne vs. Zero IF," *EETimes*, February 18, 2002.

[5] Barrett, C., "Fractional/Integer-N PLL Basics," Texas Instruments Technical Brief SWRA029, 1999.

CHAPTER 3
Cost of Test

3.1 Introduction

Cost of test is of paramount importance to semiconductor manufacturers, and ATE vendors play a vital role in combating the increasing cost-of-test pressures of the market place. Sometimes the term *cost of ownership* (COO) is used instead of *cost of test* (COT), but they are synonymous and both refer to the overall cost of test. This chapter will discuss the many aspects associated with cost of test with particular emphasis on system-on-a-chip (SOC) COT. We will briefly discuss the evolution of wafer processing as it relates to cost of test and demonstrate the orders of magnitude savings that have been realized by increasing the dimensions of the wafer. Next, we will discuss how testing strategies have evolved from the initial days of using bench equipment setups (rack-and-stack solutions; see Chapter 1) for production solutions to the highly complex ATEs that are used to test today's SOC devices. We will then discuss how the SOC has created a paradigm shift, which in many cases has invalidated the old COT models that have been in use for many years. Next, the differences between an IC device manufacturer (IDM) and a subcontract manufacturer (SCM) will be highlighted, and we will show some of the effects this has on cost-of-test modeling. We will then shift our focus into the details of cost-of-test modeling and discuss the key parameters and provide some rules of thumb that can help in determining a COT. Certain parameters are more heavily influenced by ATE than others, and examples will be provided emphasizing those parameters [1].

3.2 Wafer Processing Improves Cost of Test

There is ever-increasing and continuous pressure on semiconductor chip suppliers to reduce their overall costs. One to two decades ago, the cost of silicon manufacturing was the number one cost contributor (of which wafer production was a large percentage). Wafer diameters were 3" (76.2 mm) and smaller, design sizes were comparatively large, and process controls were not as well defined as they are today, which meant lower yields, so chip manufacturers did not get many devices from a single wafer. At that time test cost was a fraction of a percent of the overall wafer-processing chip cost; thus, high cost-of-test issues were often overlooked because even a high cost of test was in the range of 1% of the overall wafer-processing cost.

As a result, whenever a supplier wanted to test in mass production, they would simply take the solution that was used in the lab by the design engineers and duplicate that solution on the test floor. This method, although rudimentary, is justifiable as long as the cost of test never appears as a significant factor in the overall cost of the chip.

However, wafer-fabrication companies were frantically building bigger wafer-processing and -handling facilities to reduce the silicon manufacturing costs. Increasing the size of the wafers provided orders of magnitude of improvement, and these improvements were enjoyed across the entire manufacturing process. As technology progressed, wafer production became more and more streamlined and continually migrated to bigger and bigger wafer diameters.

The manufacturing savings that are realized through larger wafers is best demonstrated by example. If bigger diameter wafers could be produced using the current wafer-processing line, then more chips per single wafer could be had, thus reducing the chip cost. For digital microprocessor applications there is an added benefit from Moore's Law that each new generation of chips shrinks itself in size. So, in effect, the manufacturer is getting multiplicative factors of increasing chip volumes without increasing cost by substantial amounts. For SOC applications that have integrated digital microprocessors or other integrated digital applications, the SOC application may also benefit from Moore's Law (albeit with less effectiveness, since the digital circuitry is some percentage of the entire SOC).[1] Thus, wafers are bigger, so there are more chips per wafer, and additionally the process dimensions continually shrink, giving the manufacturer even more chips from the same wafer.

As an example, let's compare a 6" (152.4 mm) wafer versus an 8" (203.2 mm) wafer. The percentage increase in the area is determined by

$$\%(increase) = 100\% \frac{Area_{final}}{Area_{initial}} = 100\% \frac{Area_8}{Area_6} = \frac{r_8^2}{r_6^2} = 100\% \frac{16}{9} = 78\% \qquad (3.1)$$

So, increasing the wafer dimensions slightly (by 2") nearly doubles the effective area of the wafer.

Now, let's take a look at how the cost of test is positively affected when the wafer dimension increases from 6" (152.4 mm) to 8" (203.2 mm) in conjunction with the additional benefit of Moore's Law (for digital circuitry) and reduced feature size and operating voltages, which has resulted in reduced chip area. As an example, let's assume a particular die size has shrunk by 70% (from 12 × 12 mm to 7 × 7 mm, for example). The number of 12-mm² devices that can be realized on a 6" (152.4 mm) wafer is 506 (not really, because near the edges, you can't get a complete die):

$$N_{devices} = \frac{Area_{wafer}}{Area_{die}} = \frac{Area_{wafer}}{Area_{12\,mm^2}} = \frac{\pi r^2}{lw} = \frac{\pi(152.4)^2}{144} = 506 \qquad (3.2)$$

The number of 7 × 7-mm devices that can be produced on an 8" (203.2 mm) wafer is 2,647:

---

1. Although wireless chips do not follow the observations noted in Moore's Law (i.e., wireless chip transistor counts have not doubled every 18 months), the trend toward reduced feature size and operating voltage has resulted in the reduction of the chip area required to achieve specific wireless functions (Mike Golio, Golio Consulting, Personal communication to author, September 20, 2003).

$$\text{Number of devices} = \frac{\text{Area}_{wafer}}{\text{Area}_{die}} = \frac{\text{Area}_{wafer}}{\text{Area}_{7mm}} = \frac{\pi r^2}{lw} = \frac{\pi (2032)^2}{49} = 2{,}647 \qquad (3.3)$$

More than 5 times as many devices are produced on an 8" wafer versus a 6" wafer in conjunction with shrinking die sizes. This directly impacts the cost of test and has the potential to lower it by as much as 500%. In reality, what tends to happen is that more features are added to the chip so that the chip size can even increase over time. However, the kinds of functions that get tested may also change as the chip functions change. If lower-level tests can be eliminated, then there is a reduced cost of test. But, the added functionality may demand higher-level testing, which is difficult and expensive, so that the cost of test is also increased.[2] The question is, Is the decrease more than the increase? As an example demonstrating lower-level functional-testing costs versus higher-level functional-testing costs, consider a simple RF low noise amplifier (LNA) versus a multiband SOC radio. RF power is the staple of analysis used to test traditional RF devices like an LNA, and power measurements are still mainly used to test a highly integrated multiband SOC radio. Unfortunately, many more power measurements are required to test a multiband SOC radio versus the simple LNA; thus, even if the die sizes and manufacturing costs are approximately equal for the two, the cost of test is increasing for a multiband SOC radio. A higher-level functional test, like error vector magnitude (EVM; discussed in Chapter 5), could be used to test a multiband SOC radio, thus eliminating many of the traditional lower-level power tests and reducing the test cost. (Note, this is not the case with current SOC manufacturers [2]. In the extreme case, the lower-level tests and the higher-level tests are both being performed.)

Still, this took some time to mature. Building bigger wafer-producing facilities is not a simple task by any means. Tuning new processes could sometimes take a year or more, and new chip designs must go through many design iterations before they meet all of their specifications and are ready for mass production.

However, the constant demand for newer technologies drives this process, and it does not look as if this trend towards larger wafers and smaller process dimensions will slow down anytime soon (although there is a practical limit to silicon wafer dimensions).[3]

There are 12" wafer lines turning on, which offer 16 times more area than 3" wafers. In addition, device dimensions continue to shrink, thus affording faster time to market, in effect providing orders of magnitude of improvement. With larger-dimensioned wafers also comes increased processing cost. The processing costs for wafers are, of course, highly dependent on the wafer fab itself and are a source of competition between manufacturers. As a gross estimate, we can use approximate values of U.S. $5,000 for an 8" wafer and U.S. $9,000 for a 12" wafer. Table 3.1 shows the die-per-wafer increase that is realized with increasing wafer dimensions.

---

2. Mike Golio, Golio Consulting, personal communication to author, September 2003.
3. Typically, SOCs have lower volumes compared to DRAMs or MPUs. So, it is possible that due to economic reasons, many SOCs will not move to the more advanced processes. For example, mask costs for the next generation technology node are being estimated at a few million U.S. dollars (Rudy Garcia, NPTest, personal communication to author, September 27, 2003).

Table 3.1 Chip Volume Increases with Wafer Dimensions

Wafer Diameter (in)	Wafer Diameter (mm)	Die per Wafer	Increase from Previous Size (%)
3	75	44	
6	150	176	400
8	200	314	78
12	300	706	25

The table is normalized to a device that is 100 mm^2 and does not consider the added benefit from shrinking technology. Clearly, as a minimum, a 12" wafer-fabrication facility can offer a 25% reduction in manufacturing costs versus an 8" wafer fab.

Take a look at Figure 3.1, which shows the trends in manufacturing versus test cost that have been realized in part by larger wafer-processing facilities.[4] The silicon manufacturing cost (of which the wafer-processing cost is a large percentage) has steadily been dropping, while the test cost has remained relatively flat or, in some instances, has risen slightly.

This means that the once overlooked cost of test is now something of paramount importance. Currently, for many of today's SOC manufacturers, cost of test is in the range of 3% to 30 % of overall manufacturing costs versus the fraction of a percent that it once was [3].

3.3 Early Testing of the SOC

The system-on-a-chip (SOC) chipsets compound the cost-of-test problem. Because test costs were historically a much smaller percentage of overall manufacturing

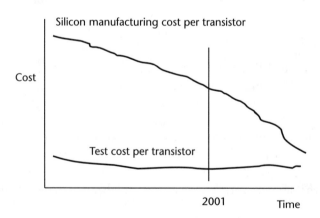

Figure 3.1 Manufacturing and test cost trends.

---

4. Figure 3.1 only deals with the processing costs and ignores the potentially much larger depreciation costs of implementing a new fab. It also ignores the wafer sort yield of these devices (Rudy Garcia, NPTest, personal communication to author, September 27, 2003).

costs, this allowed test methodologies and strategies to evolve without much attention or standardization. A manufacturer that is testing RF front-end power amplifiers has completely different needs than does a memory chip or DSP manufacturer. Thus, it should be no surprise that there are entire ATE companies that have existed only to build test solutions for a particular type of chip. For example, there are companies that historically have only built logic test systems to test logic devices, and there are other companies that have built only digital test systems to test strictly digital devices. This method supported market demands and expectations until a few years ago, when the industry started seeing multiple technologies merge onto a single chip, which was quickly coined SOC. Then, suddenly, no test equipment supplier offered a single complete solution to meet all of the test needs. This had an overall negative impact on the cost of test. In some cases, manufacturers were testing part of the chip's functionality on one type of tester and then testing the chip's remaining functionality on a second tester (this more than doubled the COT) [4].

SOC testing created a market that ATE companies are rushing to fill. The big issue has been that most of the ATE companies have been in existence for tens of years, and their specialized technologies are based upon years of experience in a particular testing discipline. It is extremely difficult to merge a completely new technology into an existing infrastructure and have the new merged technology meet the same performance and quality standards that the core technology offered.

3.4 SCM and IDM

An integrated device manufacturer (IDM) is a company that offers all of the necessary processes to generate new devices (i.e., design, layout, wafer fabrication, packaging, testing, and so forth). A subcontract manufacturer (SCM) specializes in a particular area of the silicon process (these are either in the category of design houses or test houses).

Outsourcing business aspects that are not part of your core competency is a very useful way to reduce COT. As COT became a larger percentage of the manufacturing cost, some silicon IDMs (whose core competency was in the wafer-fabrication process) decided to outsource testing. SCMs specializing in testing (test houses whose core competency was testing) emerged offering test services and tester rates by the hour (see Section 3.6.3). SCMs purchase ATE from multiple ATE vendors and then offer hourly rates. This has the effect of acutely highlighting COT differences among the various ATE companies (two major factors being acquisition cost and test time, which will be discussed later).

3.5 SOC Cost-of-Test Paradigm Shift

The rapid introduction of successive generations of SOCs, each exhibiting higher levels of performance and integrating more digital, analog, and RF functions, has created a new range of innovative and affordable products. In the process, prices have plummeted. Bluetooth SOC devices, for example, were marketed to require a manufacturing cost of $5, before consumer acceptance could be widely achieved [5]. This cost entry point has already been achieved and surpassed by many manufacturers, but to do so, the testing methodologies were forced to become COT centric. Although low prices are beneficial for the consumer, unfortunately for the SOC manufacturer, this has given the SOC business three characteristics unique to this segment of the semiconductor industry [6]:

1. Unpredictable market demands;
2. Shrinking product life cycles;
3. Unrelenting cost pressures.

The SOC manufacturer must continually innovate to remain competitive. Thus, they design highly complex devices on leading-edge processes that provide the best cost, performance, and feature sets [6].

The market unpredictability makes it impossible for the SOC manufacturer to predict which designs will win in the market place and which will lose. In many cases, the volumes never materialize, and those costs must be absorbed by a more successful SOC, further increasing the COT on the winner.

Historically, semiconductor devices had a life span of 1 or more years with significant volume, but the SOC often has a significant volume measured only in months, before being replaced by the next emerging SOC device.

This short life span makes it much more difficult to recover the heavy investments that were required in the first place, because the device undergoes severe price erosion shortly after introduction [6].

These three characteristics have been significant factors in increasing the cost of test that now is sometimes as high as 25% or more of the overall manufacturing costs (Figure 3.1) and is causing a paradigm shift in the traditional COT model. Purchasers of ATE equipment historically depreciated traditional ATE equipment costs over 5 years, but with continuous innovation in SOC devices, current ATE in some instances no longer has a life span of 5 years. Additionally, traditional ATE may not be fully equipped to handle all of SOC testing requirements, so the equipment's utilization is poor. Many more upgrades can be required to keep up with the innovative trend of SOC devices, which further drives up the COT. Test-development costs also increase because multiple testing disciplines (digital, mixed-signal, and RF) are required to create a test solution (i.e., more engineers are needed). Thus, the ATE industry is under tremendous pressure to find innovative ways to lower the cost of test. To do that, new innovative test techniques and testing methodologies are being developed that can be measured and compared by using key cost-of-test modeling parameters.

3.6 Key Cost-of-Test Modeling Parameters

Trying to model the COT of a test cell requires a fairly large number of parameters and can be a demanding task [6]. The key parameters include:

- Test time;
- ATE equipment cost (capital cost, fixed cost);
- Lifetime;

- Utilization (flexibility, upgradeability, availability);
- Yield;
- Measurement accuracy (repeatability);
- Space (floor space);
- Maintenance (spare kits, support contracts, calibration).

A simple expression relating the key parameters to COT is

$$COT = \frac{(Fixed\ cost + Recurring\ cost)}{(Lifetime \times Yield \times Utilization \times Throughput)} \tag{3.4}$$

It is also interesting to note that there is a cost of ownership (COO) standard defined by SEMI as E.35, which provides an extensive definition for COT/COO that has over 150 input variables. However, this standard is difficult to utilize without already knowing many of the variables (which is often the case), and the standard is currently being revised.

Establishing exact values for (3.4) can be very time-consuming and expensive. Manufacturers have sophisticated process-tracking tools that track COT parameters that are critical to their individual models. Moreover, many of the parameters, like lifetime, yield, utilization, and throughput, are really only known after the life of the SOC. To make matters worse, SOC devices have shorter life cycles than other traditional devices. Models based on (3.4) are typically used in a more qualitative manner or to track perturbations on existing costs, rather than as exact evaluation tools. For these reasons, it is important to develop rules of thumb for many of the parameters and a general understanding of how the factors affect COT. With that in mind, let's take a look at each individual parameter to determine what assumptions, if any, can be made to make it easier.

3.6.1 Fixed Cost

The fixed cost consists of the capital equipment cost and floor-space cost. With typical ATE equipment sometimes having acquisition prices well over $1 million, capital cost is usually one of the dominant factors still used to determine the COT. However, the importance of capital cost is often supplanted by the higher throughput that a more expensive ATE solution offers and should not be easily disregarded when evaluating COT.

3.6.2 Recurring Cost

Recurring cost can be defined as things like calibration, cleaning, general maintenance, support contract costs, and software subscription costs. Although each of these items is indeed a cost, it can be argued that any ATE purchase carries these costs and that their contribution to the overall COT is essentially the same for different ATE vendors. For instance, recurring costs are often taken to be 10% of the ATE acquisition cost [6]. If we accept that argument, then we can remove this parameter or normalize it out of the COT equation to obtain a COT model where the numerator only depends on the fixed cost of the ATE equipment, as shown in the following equation:

$$COT = \frac{Fixed\ cost}{(Lifetime \times Yield \times Utilization \times Throughput)} \quad (3.5)$$

3.6.3 Lifetime

The lifetime of ATE equipment has historically been depreciated over 5 years, but if the ATE is inflexible, then perhaps a 3-year depreciation is called for. What does depreciating across 3 years do to the COT? If we take a simple COT example [6] and assume a $1 million acquisition price for the ATE, the test cost per hour is $114 per hour [5]. (Note that this assumes 100% utilization and 100% throughput.)

$$COT = \frac{1{,}000{,}000}{365\left(\frac{days}{year}\right)24\left(\frac{hours}{day}\right)} = \$114/hour \quad (3.6)$$

In reality, the COT will be higher depending on the test yield, equipment utilization, handler costs, labor costs, and so forth. Continuing with (3.6), the COT per second of test time is calculated to be approximately 3 cents. It is more common to use 5 cents per second because the reality is that there are many other COT variables that are dynamic and not easily determined. One method to combat a decreasing depreciation period is to lengthen the useable life of the equipment. However, to lengthen the useable life, the ATE must be highly flexible and adaptable to the SOC market. The emergence of SOC platforms is one method by which ATE vendors are addressing the highly sensitive COT nature of the SOC market. The SOC platform is intended to lengthen the useful life of the ATE by offering IDMs and SCMs the ability to configure the equipment dynamically in a matter of hours to the testing requirements of the SOC. As long as the SOC platform can be dynamically configured and easily upgraded, the equipment has a longer useful life (5-year depreciation is justified) that can dramatically reduce the COT.

3.6.4 Utilization

Utilization is the percentage of time that the test equipment is actually being used for production testing. Repairs, maintenance, calibration cycles, test development, and basically anything other than production testing, are excluded. In the ideal case, the test equipment would be testing parts 24/7; however, in reality, even world-class manufacturers of high-volume devices have difficulty reaching utilization rates greater than 90% [6]. It is not uncommon to see ATE sitting idle on an SOC test floor while devices are waiting in queues because the tester does not have the right configuration [6]. If we return to the $114/hour rate, we can view the impact that utilization has on the COT. For every hour or day the ATE sits idle, it costs the manufacturer $114/hour, or $2,736/day. A 50% utilization rate means the COT increases to $228/hour or $5,472/day! These values are really just a measure of the absolute minimum COT to the manufacturer. In reality, the lost-opportunity cost and slower time to market due to low utilization rates can easily increase the COT by orders of magnitude. For example, a test time of 1.5 seconds with a device price of $2.00 would mean an extra $4,800 of revenue is not realized for every single hour the ATE is sitting idle. This is nearly 50 times greater than the depreciation cost! To help silicon manufacturers better control their utilization rates, ATE manufacturers are offering flexible, scalable platforms with short calibration cycles and low mean time between failures (MTBF) and mean time to repair (MTTR). Utilization is heavily dependent on the flexibility of the SOC ATE, as well as on the silicon manufacturer's management of his test floor. If the SOC ATE is flexible enough to allow dynamic configuration on the test floor, this can provide enormous COT savings by increasing the utilization of the ATE.

For the sake of argument, assume that an SOC is sufficiently flexible and dynamically configurable and that the test floor is perfectly managed so that it enjoys a 100% utilization rate. Even if the argument is rejected, it can at least be argued that the utilization burden is shared by the ATE vendor and the management of the test floor because the ATE must be capable of being easily utilized, but the test floor must keep the ATE loaded and testing. Additionally, assume that the lifetime has been sufficiently extended so not to be a critical factor in the COT equation.

Given that scenario, the utilization and lifetime parameters can be normalized out of the COT equation to provide

$$COT = \frac{Fixed\ cost}{(Yield \times Throughput)} \quad (3.7)$$

3.6.5 Yield

An entire chapter could be written on the subject of yield. We will attempt to point out the most significant aspects of yield and its affect on COT. A lower bound on the COT can be calculated by assuming the COT due to yield losses is equivalent to the scrap costs. However, since SOC devices have such a short life cycle, it would be more accurate to treat yield loss as lost revenue [6]. To better comprehend lost revenue, let's take a look at the two types of errors that impact yield, Type I, and Type II errors.

1. A Type I error (also called false negative, or alpha error) occurs when the ATE fails a good part. The part is then thrown away or scrapped. This results in building more wafers to achieve a specific volume and longer rental rates at SCMs [6]. The COT incurred from Type I errors can be lower bounded as the scrap cost.
2. A Type II error (false positive or beta error) occurs when the ATE passes a bad part. This means a bad part makes it through the screening process and into the customer's product. Type II errors occur due to insufficient guard banding, can be more costly than Type I errors, and can have drastic consequences [6]. Shipping unacceptable devices risks having the device designed out of a product [6]. It is extremely difficult to quantify the COT due to a Type II error. A crude rule of thumb is that a Type II error costs 10 to 100 times more than a Type I error.

With sufficient guard banding, Type II errors can be completely eliminated, but at the expense of increasing the Type I errors. Let's assume this scenario to be the standard, in which case minimizing the necessary guard banding is the key underlining element of yield. To minimize the guard banding, the accuracy of the ATE must be maximized (unfortunately, maximizing accuracy often leads to longer test times, which increases COT, so a trade-off must be determined).

Appropriate guard banding implies that the correct performance specs are implemented. Although this seems obvious, an improperly established spec can wreck havoc on yield. Consider this real example (although the absolute values have been changed to protect privacy), where a device has a third-order intercept (TOI) spec of 15 dBm. Previously, the device was characterized and known to have a TOI range of 15 to 18 dBm. The ATE is set to pass devices having higher than 15 dBm and lower than 20 dBm (the 20 dBm is used as a sanity check). The wafer processing shifts slightly over time such that the devices' third-order tones drop into the noise floor of the ATE. This has the effect that the TOI is improved dramatically, but the ATE yield drops off as it starts failing good parts, when in fact all the parts are good parts and should pass. More than 4 days were spent running correlation lots to isolate the cause of the yield drop-off. Using our conservative lost-opportunity cost of $4,800/hour implies that more than $450,000 dollars were lost when, in fact, there was never a problem.

3.6.6 Accuracy as It Relates to Yield

Guard bands are required due to inaccuracies of the ATE. A more accurate SOC ATE will have a higher yield than a less accurate ATE. How accuracy (ATE specifications) affects yield is difficult to translate directly, but it can be reasonably estimated from prior-generation devices [6]. Tester accuracy is often undervalued in the COT model, because it is so buried in the details and difficult to quantify.

Consider the following memory testing model from [7] to illustrate the point. DRAM yield impact versus tester accuracy for improving picosecond edge placements was generated for Table 3.2.

Assuming a price premium of $5 between devices binned at 800 Mbps versus 600 Mbps, the 95% ideal yield case shows a 14% increase in the yield (69.2% to 83.3%), where the overall timing accuracy (OTA) of the ATE was improved by 60 ps. For 100 million parts produced annually, savings of 14% in yield results in savings of $70 million (i.e., $1.1 million for every extra picosecond of accuracy) [7]. If the tester tests 10 million parts a year, that translates to a $7 million savings, more than paying for itself in less than 1 year [7].

If this is the case, then why are IDMs and SCMs not eagerly purchasing an ATE for twice the capital cost amount, if it can save them $7 million in 1 year? The answer is often risk aversion; that is, an ATE purchaser would rather risk buying five cheaper ATEs with different configurations knowing that one or more of them might sit idle versus one expensive ATE, which when being properly utilized is of course *saving* a lot of money, but when sitting idle, is costing a fortune due to the

Table 3.2 Actual Yield Versus Tester Accuracy

Ideal Yield (%)	OTA = ±160 ps	OTA = ±100 ps	OTA = ±80 ps	OTA = ±60 ps
60	51.67	55.3	56.4	57.5
75	53.3	63	66	68.7
95	69.2	83.3	86.6	89.6

(OTA = overall timing accuracy)
Source: [3].

3.6 Key Cost-of-Test Modeling Parameters

higher fixed cost.[5] Purchasing multiple less-expensive ATEs also makes it easier for the ATE purchaser to scale his test-capacity needs. If the manufacturer is utilizing the more expensive ATE strategy and needs slightly more test capacity, he is forced to purchase another expensive ATE, which may not be fully utilized, thus detracting from his COT savings.[5] Another point to consider is that in many testing situations, the accuracy has no effect on yield, whereas in other testing situations, the accuracy is of paramount importance. Consider Figure 3.2(a, b). In Figure 3.2(a), the ATE is highly inaccurate, but the measurement result is so far above the specification that even with guard banding included to compensate for the inaccuracy, there is no depreciable impact to the yield. Conversely, Figure 3.2(b) shows an ATE with higher accuracy, but because the measurement result falls well within the Gaussian distribution, there is a noticeable negative impact on the yield. This is one of the main reasons it is so difficult to quantify the impact of accuracy on yield. If the

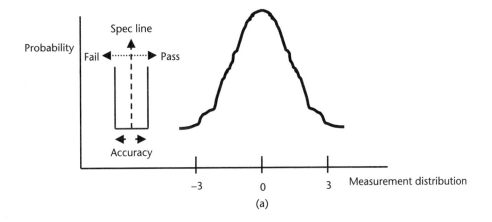

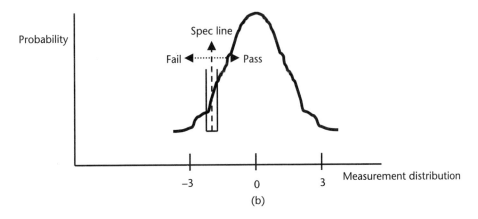

Figure 3.2 Impact of accuracy on yield: (a) low accuracy, low impact; and (b) high accuracy, high impact.

---

5. John McLaughlin, Agilent Technologies, personal communication to author, August 2003.

device is perfectly manufactured, then high inaccuracies of the ATE are negligible, whereas if the device is poorly manufactured, then smaller inaccuracies can have an enormous adverse effect on the yield. If the yield fallout is determined to be due to only one or two test items or specifications, then the testing strategy can be focused and the test cost reduced. If, however, the yield fallout is due to a wide range of test items, then increased testing or more accuracy may be required, and the test cost is increased. Both cases are common in industry. For traditional devices, the testing strategy evolves over the life of the device, and many nonessential tests are removed (see Section 1.2); still others are improved to reduce test time, and the test cost is continually reduced. However, for SOC devices that have a shorter life cycle, it is less likely that such benefits can be had, in which case accuracy can become a dominant COT factor.

Therefore, again, it can be argued that yield is a shared COT element between the ATE and device manufacturer. The ATE vendor has no control of the wafer-fabrication process, and the SCM or IDM has no control over the accuracy of the ATE.

For the moment, accept a very bold and completely unsupported statement, that yield is more heavily impacted by the fabrication process than by tester accuracy.[6] Or, to state it differently, let's assume that the device yield is always 100% and that the ATE has no inaccuracies. We are doing this only to simplify the COT equation to

$$COT = \frac{Fixed\ cost}{Throughput} \qquad (3.8)$$

where now the COT only depends on two factors, fixed cost and throughput. Recall the assumptions that were made to reduce (3.4) to (3.8). The assumptions of who more heavily influences a given COT parameter are shown in Table 3.3.

As a final example, let's examine two different COT scenarios using (3.8):

Scenario 1

- ATE capital cost is $1 million.
- Test time is 4 seconds per SOC device (throughput is .25 devices/sec).

Table 3.3 Influence on Key COT Parameters

COT Parameter	ATE	IDM/SCM
Fixed cost	*	
Test time	*	
Accuracy	*	
Utilization		*
Yield		*
Space	*	
Maintenance		*
Labor		*

6. This can also be thought of in terms of device specification versus process capability/variation. Many of the defects found in the sub 0.18 μm are not "stuck at," but are more parametric in nature (Rudy Garcia, NPTest, personal communication to author, September 27, 2003).

Scenario 2

- ATE capital cost is $2 million.
- Test time is 2 seconds per SOC device (throughput is .5 devices/sec).

In scenario 1 the capital cost is lower, but the test time is twice as long, while in scenario 2, the test time is lower, but the capital cost is twice as much. However, both scenarios have an equivalent COT at 2.5 cents/s. The equivalent COT of the two scenarios implies that either test solution can be chosen, resulting in the same COT. This result, while true in theory, is somewhat misleading. There are approximately 31.5 million seconds per year of available test time, and the object is to maximize the throughput. Another important point to consider is the number of devices that can be tested. For scenario 1, approximately 5.1 million parts per year can be tested, whereas for scenario 2, approximately 10.2 million parts per year can be tested. Depending on the profit earned from each device (say $1), the difference can be quite substantial (e.g., $5.1 million), which, after subtracting the extra capital cost of scenario 2, leaves a total of $4.1 million. One could argue that the same benefit is realized by purchasing two testers at $1 million each (i.e., twice scenario one). However, there would be many additional costs with the second ATE (floor space, handler, repairs, calibration, maintenance, labor, and so forth), so that in fact it is apparent that scenario 2 would be the better choice. In reality, however, risk aversion can nullify the extra savings that could be had because again ATE purchasers would rather risk having one of the less expensive ATEs sitting idle and one generating some revenue versus risk having the more expensive ATE sitting idle with no revenue being generated.[5]

3.7 Other Factors Influencing COT

3.7.1 Multisite and Parallel Testing

The SOC brings with it increasing cost-of-test reduction pressures that often cannot be met with single-site testing. Multisite testing is another method to reduce the cost of test. Testing devices in parallel increases the throughput considerably and thus lowers the COT. For example, if two SOC devices can be tested in parallel, theoretically the throughput can be doubled or, equivalently, the COT halved. Of course, handler capabilities and load board space must be considered along with the quantity of testing within a test plan that can be implemented in parallel from a multisite implementation. However, for large volumes, the savings that can be realized are hard to ignore. There are two basic types of multisite testing: (1) parallel, and (2) concurrent [8, 3, 9]. Parallel and concurrent testing of wireless SOC devices are discussed in detail in Chapter 7, along with the COT considerations associated with each. There is a multitude of parallel/concurrent implementations that can be had, especially when one considers that each wireless SOC usually has multiband radios. In such instances, if the ATE software and hardware are flexible and sophisticated enough, heterogeneous parallel (i.e., testing the TX section of device 1, while simultaneously testing the RX section of device 2, and then reversing the procedure), heterogeneous concurrent, and heterogeneous parallel/concurrent combinations are possible, which take advantage of the SOC and ATE architectures.

3.7.2 Test Engineer Skill

Ask any SOC engineer, and he will tell you the complexities of the job have risen by at least a factor of the number of technologies, that is, three, that have been integrated into SOC devices. There is a good reason that digital, mixed-signal, and RF engineering are classified as disciplines. Each discipline requires not only a broad understanding of the technology, but to be truly knowledgeable in any one field also requires a tremendous depth of knowledge within that discipline. SOC devices bring new challenges to the test/product/application engineer. The engineer must suddenly be able to understand and test aspects of the chip that are completely foreign to him. Indeed, this new breed of SOC engineer must be able to accommodate logic, memory, microprocessors, microcontrollers, digital, DSP cores, mixed-signal, and of course RF into his testing solutions. Therefore, the skill-level mastery or deficiency of each of the disciplines can greatly decrease or increase the COT. The engineer's skill level can drastically impact the test-solution schedule (time to market) and, of course, test time (throughput). With SOC devices having such a short life cycle, late market introduction can seriously erode the manufacturer's profit taking, and if the test time is not properly optimized, the resultant lower throughput (higher test time) will also increase the COT and detract from the manufacturer's profits [10].

3.8 Summary

An introduction to cost of test was discussed beginning with an emphasis on manufacturing costs that have largely been addressed with increasing wafer dimensions. This helped drive down the manufacturing cost to within the same COT magnitude range, thus underlining the need to address the rising COT. The SOC and SCM/IDM impacts on COT were then discussed. A simple COT model was offered, and the key parameters were examined in detail. Assumptions and arguments were presented to reduce or isolate certain key parameters so that conclusions and decision criteria for evaluating the COT of a test solution could be determined.

References

[1] Newsom, Tom, "SOC Test Challenges for the New Millennium," *IEEE International Test Conference,* Baltimore, MD, November 2001.

[2] Lowery III, Edwin, "Integrated Cellular Transceivers: Challenging Traditional Test Philosophies," *IEMT,* San Francisco, CA, July 16–18, 2003.

[3] Fischer, Martin, "Concurrent Test—A Breakthrough Approach for Test Cost Reduction," *Semicon Europe 2000,* at www.ra.informatik.uni-stuttgart.de/~rainer/Literatur/Online/T/15/3.pdf.

[4] Lecklider, Tom, "Reducing the Cost of Test in 2003," *Evaluation Engineering,* at www.evaluationengineering.com/archive/articles/0803icate.htm, accessed 2003.

[5] "$5 Silicon: No Way or Yes Way," at www.csr.com/enews/hw001.html.

[6] Garcia, Rudy, "Redefining Cost of Test in an SOC World," *Evaluation Engineering*, at www.evaluationengineering.com/archive/articles/0603ic.htm, accessed 2003.

[7] Dalal, Wajih, and Miao, Song, "The Value of Tester Accuracy," *IEEE International Test Conference*, Atlantic City, NJ, September 1999.

[8] Mast, K., "Reducing Cost of Test Through DFT," Agilent Technologies application note, 2003.

[9] Hilliges, Klaus-Dieter, et al., "Test Resource Partitioning for Concurrent Test," *IEEE International Test Conference*, Baltimore, MD, November 1–2, 2001.

[10] Burns, M., and Roberts, G. W., *An Introduction to Mixed-Signal IC Test and Measurement,* Oxford: Oxford University Press, 2001.

CHAPTER 4
Production Testing of RF Devices

4.1 Introduction

Descriptions and examples of traditional RF production testing will be presented in this chapter. Each section throughout this chapter will build on the previous sections. Most of the traditional measurements presented in this chapter are still utilized today to determine the quality of RF SOC devices before they are shipped to the end customers (wireless phone and consumer device manufacturers). Nearly every traditional measurement is in some way connected to the basic "power measurement," so detailed descriptions and definitions of the various forms of power are provided as a foundation to the reader. The remaining traditional RF measurements will be described from the perspective that the reader has a good understanding of making power measurements.

Many of today's RF measurements, whether they are performed on an RF-centric or SOC-centric device, have their roots tied to RF power. As such, it is useful to understand a little of the background and history of how these measurements were developed. For example, spurious tones, harmonics, third-order intercept, power compression, and adjacent channel power are all forms of RF power measurements. Additionally, if the input and output impedance matching of a device under test (DUT) are close to ideal, gain can also be considered a power measurement. In one form or another, it has always been necessary to determine the output power level of RF devices.

4.2 Measuring Voltage Versus Measuring Power

Why not just measure voltage and current? Ohm's Law is the linear estimate of Maxwell's electromagnetic field equations. Maxwell's equations show that voltage and current are actually waves dependent on frequency. Think about measuring a voltage sine wave with an oscilloscope. The location of the observation point determines the actual voltage that is measured. If the time base of the oscilloscope is continually reduced, eventually a straight line will be seen on the display because only a very tiny piece of the wavelength is being examined. This analogy can be applied to an RF signal that is being measured on the same oscilloscope. Instead of continually being reduced, the time base of the oscilloscope is held constant, and the frequency of the signal is continually reduced. The same thing happens on the display. At first, we can see a sine wave with multiple periods, but eventually we see only a straight line across the display. If we see a straight line on the display and ask the question,

What is the voltage? we can simply use Ohm's Law, because at very low frequencies, the wavelength is immaterial. The higher the frequency is, the less meaning Ohm's Law has. So, for frequencies starting at about 30 MHz and higher, power measurements become more important.

4.3 Transmission Line Theory Versus Lumped-Element Analysis

Those readers who are not familiar with microwave circuit analysis may not be aware of the physics that lies beneath the simple equation of Ohm's Law ($R = V/I$). Ohm's Law is derived from, or more accurately, is a simplification of, Maxwell's equations. Without getting into a detailed analysis, Ohm's Law is derived from Maxwell's equations based on some assumptions. Namely, those assumptions are the mathematical boundary conditions that state that frequency is assumed to be very low (relative to the entire frequency spectrum).

Measurement of a voltage at dc or low frequency (less than approximately 30 MHz) is a straightforward task using a handheld multimeter or oscilloscope. Measuring voltage levels at frequencies higher than that becomes a more arduous task. Measurement of voltage, or power, at high frequencies involves analyzing where the energy is as a function of position. Therefore, the measurement involves the measuring of waveforms. This provides the user with information on phase, in addition to the other parameters.

Spatial information is not used in low-frequency measurements because the wavelength of the electrical signal is much larger than the device that is being measured. For example, the wavelength of a signal at 100 Hz is 3,000m long. This is obviously much longer than the capacitor or inductor (which is, say, 10 mm in length) that the signal will pass through. From the time the signal has entered the device to the time it leaves the device, essentially no difference in the signal's phase is seen from the device's point of view (actually, there is some difference, but mathematically this assumption can be made). This is depicted in Figure 4.1 (a, b) for a long wavelength and short wavelength.

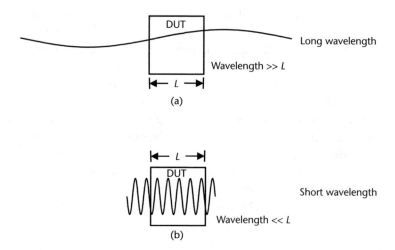

Figure 4.1 Effect of wavelength on DUT.

Also shown is the other extreme, a signal operating at 100 GHz has a wavelength of 0.5 mm (in freespace). From the time the signal enters a device to the time it leaves the device, the device will encounter many full wavelengths. The energy at any point of time is dependent upon where the signal is measured in the device.

This leads to an understanding of how making accurate measurements at high frequencies could be a difficult task. Direct waveform measurements do involve the magnitude and phase of a wave traveling in a given direction or of a standing wave. However, for most production RF testing, the magnitude of power is the most important parameter and will be the focus of this chapter.

4.4 The History of Power Measurements

In the late 1930s, power was initially measured with a fluorescent screen. Sigurd and Russel Varian were pioneers of this method when they were developing a method to measure the power of their oscillating klystron cavity. The oscillating frequency of the klystron was too high to measure it directly, so they drilled a hole in the cavity and placed a fluorescent screen alongside it. If the klystron was in oscillation, then they could tell by the fluorescent screen lighting up. Not only that, but additionally they could also use the screen to provide a gross indication of the power level. A higher power level is indicated with a brighter fluorescent screen. A lower power level produces less brightness.

World War I greatly enhanced the rate of technological advancements, especially in the area of RF. Long-distance communications, radar, pulse jamming, and receiver enhancements all benefited from the war. High power levels were difficult to measure with early long-distance communication systems requiring 100 to 1,000W. This is a far cry from the mobile phones of today that require in the range of a few milliwatts to just a few watts. There are two key things to consider when attempting to measure high power. The first key thing to consider is heat dissipation. Somehow the thermal energy from the RF signal must be absorbed. The second, and perhaps more important, issue is the termination of the measuring device. If the signal is not terminated properly, all of the power can be reflected back, which can damage or destroy the device or test equipment. Early high-power measurements were made by terminating the output of the device into a material that could absorb the RF energy (water, for example). The measurement was then made by measuring the heat buildup over time. The power could then be determined by dividing the total heat energy divided by the time.

Again, another war spurred the RF and wireless communications field. During World War II, detectors were improved and performed to higher power levels and higher frequencies. Comparison techniques were improved so that power levels measured using these detectors could be measured against known values.

The important point to remember is that during each stage of development of the above scenarios, a reference power level was used to determine the actual power level measured. Russel and Sigurd Varian used a fluorescent screen, and by watching the glow of the screen, they could determine if the power level was increasing or decreasing. The method was rudimentary at best, but it allowed them to estimate the power level. If it is known how much energy it takes to heat up a certain amount of water from room temperature to its boiling point, then that can be used to measure the high-power output of an RF device. Lastly, if a detector is calibrated to a known RF source generator, then it is possible to use that detector to measure other unknown RF power levels [1].

4.5 The Importance of Power

As the frequency of a signal increases, power becomes a more and more important quantity. The output power of any system is often the most critical factor in the design and ultimately the purchase of nearly all radio frequency equipment. This was made apparent during the brief description of the previous wars when a system that had higher output powers, could transmit over a longer distance, thus giving the owner of that system a decisive advantage. Likeminded mobile service providers are competing in this same manner. If one provider can produce higher power than another provider, this allows him to space his base stations further apart. Thus, his coverage area increases over the competition's or his infrastructure cost decreases as compared with his competition's. An important thing to keep in mind is the concept of equity in trade. A customer has a desire to purchase a product with a specified power performance that meets his needs. The final production-line performance results should match his original sample data. In today's smaller world, design, fabrication, manufacturing, packaging, and test are scattered about the globe, and it is important that the measurements that are used to define a device's performance be consistent (see Section 1.8).

Any measurement inconsistencies or uncertainties directly impact the bottom line. Wireless communication systems are all designed with some sort of power budget in mind. The antenna is a certain size, because the area of coverage is predetermined. The base stations are a certain distance apart, because the power levels are predetermined. The battery life of the phone or handset is based upon optimum or near optimum power use levels. If one or more of those assumptions are wrong, then this can dramatically inflate the cost. This is best illustrated with an example. An amplifier that produces 10W produces twice the power of an amplifier that produces only 5W. The 10-W amplifier has a 40% more radial range versus the 5-W amplifier. It makes sense that the 10-W amplifier is more expensive and valuable than the 5-W amplifier. With the 10-W amplifier the base stations can be spaced further apart, which means less capital expenditures for hardware and infrastructure costs. Yet, if the measurement uncertainty is ±0.5 dB, then that translates to a possible 10% reduction in overall coverage area. The service providers would not be very happy to hear that instead of installing the planned 30 base stations, they need to install an extra 3 just because of measurement uncertainties in the range of 0.5 dB.

As the power level increases, the manufacturing and test costs per decibel increase as well. The extra costs come from increased complexity of design, expense of active devices, skill in manufacturing, difficulty in testing, and degree of reliability. Additionally, guard bands are designed in to prevent overstressing these costly devices. It gets even more complicated as the frequency increases. The higher the frequency becomes, the easier and indeed more necessary it is to make a direct measurement of power versus either a voltage or current measurement.

RF systems of today depend on signal formats and modulation schemes that make power measurement techniques more critical. They have to handle fast digital phase-shift-keyed modulations, wide bandwidth, multiple channel carrier signals (like OFDM used for WLAN [2], and other complex formats. These formats make measuring power more difficult by complicating the process of selecting a sensor. There is a high demand for wide-bandwidth sensors, but that comes with a price of very complicated signal-shaping circuits that are needed for these wide-bandwidth signals. It is easy to see why power then becomes one of the most important parameters for RF and wireless SOC devices [1].

4.6 Power Measurement Units and Definitions

The unit of power is the watt (W), which is established by the International System of Units (SI). One watt is one joule per second. If you are new to RF, then you might be used to thinking of power in terms of voltage or current. For example, power equals the voltage squared divided by the resistance. This is really only valid at dc or near-dc frequency levels. In fact, it is interesting to note that voltage is actually derived from the watt and not vice versa, which is a common misnomer. A volt is 1W per ampere (see Chapter 6 for many of the same definitions presented from a mixed-signal point of view).

4.7 The Decibel

The ratio of two power levels or relative power is often more desired over the absolute power. Relative power is the ratio of one power level, P, to another power level, P_{ref}. Relative power is dimensionless, as is relative anything, because both the numerator and denominator have the same units. Relative power is almost always expressed in decibels (dB).

The decibel is defined by

$$dB = 10 \log_{10}\left(\frac{P}{P_{ref}}\right) \quad (4.1)$$

The decibel is our friend and is used ubiquitously when discussing RF and wireless SOC measurements. It has two big advantages over the watt. One advantage is that the decibel compacts the numbers. This is very useful and stems from the fact that wireless systems have enormous dynamic ranges. Trying to write down everything in terms of watts would become very confusing due to the enormous orders of magnitudes that are involved. For example, a Bluetooth radio modem might have an operating range from −70 dBm to +5 dBm or 75 dB of operating range. This does not seem like a large range, and it is very easy to write it down, or graph it. But converted to watts, we get 0.1×10^{-6} and 3.16227766 respectively, which is 7 orders of magnitude of dynamic range. The second advantage is that decibels are added to and subtracted from each other when computing system gain and noise-figure performance. Watts would require multiplication and division. It is much simpler and easier to add −70 dBm to 5 dBm than it is to multiply 0.1×10^{-6} mW by 3.16227766 mW.

4.8 Power Expressed in dBm

The noise floor of wireless SOC devices continues to be reduced. A lower noise floor means smaller and smaller amplitude signals can be transmitted and received between wireless devices. These very low-level signals beg for the introduction of the term dBm. The dBm is used ubiquitously when discussing nearly every RF measurement of any RF or wireless SOC device. The formula for dBm is similar to that for decibels except that P_{ref} is predefined to be one milliwatt (1 mW).

$$P_{dBm} = 10 \log_{10}\left(\frac{P_{mW}}{1\,\text{mW}}\right) \quad (4.2)$$

The nice thing about this expression is that P_{mW} is expressed in milliwatts and is the only variable, so dBm is used to measure absolute power. A "+" sign is assumed and used to indicate dB above one milliwatt. A "−" sign indicates dB below 1 milliwatt. As an example, the Digital Enhanced Cordless Telecommunications (DECT) standard can operate down to around −100 dBm. This translates to an absolute power of 100 dB below 1 mW (i.e., the absolute power equals 1×10^{-13} W)!

4.9 Power

The term *average power* is used to specify almost all RF and wireless SOC devices. The terms *pulse power* and *modulated power* will be discussed later.

Power meters are inherently slow and are therefore not widely used in production-test solutions to measure RF power. If a power meter is not used to measure RF power, but instead the RF signal is downconverted to an IF signal, and then the IF signal is digitized across some known resistor, R, the discussion of RF power is reduced to the block diagram shown in Figure 4.2.

In basic electrical theory, power is said to be the product of voltage and current. But for ac voltage cycles, the product $V \times I$ varies during the cycle. This is shown by the P curve. Notice that the P curve varies by twice the frequency ($2f$) of the voltage cycle. The P curve has both a dc component and an ac component. When referring to power, this most commonly refers to the dc component of the power product and not the ac component.

As was mentioned earlier, the fundamental definition of power is energy per unit time. This is in agreement with the definition of a watt as energy transfer at the rate of one joule per second. When looking at the P curve, we are suddenly faced with a question. Over what time is the energy transfer rate to be averaged when computing the power? Figure 4.2 clearly shows that if a narrow time interval is chosen, it is possible to get varying answers for the energy transfer rate. What must be ensured is to take an integer number of ac periods. The power of a continuous wave (CW) signal at frequency f_0 ($1/T_0$) is defined as

$$P = \frac{1}{nT_0}\int_0^{nT_0} V_{peak}\sin\left(\frac{2\pi}{T_0}t\right) \times I_{peak}\sin\left(\frac{2\pi}{T_0}t + \phi\right)dt \quad (4.3)$$

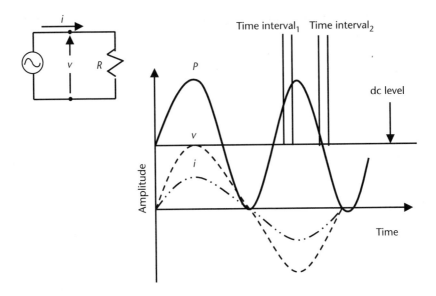

Figure 4.2 Power versus time curve.

where T_0 is the ac period, V_{peak} and I_{peak} are the peak values of the voltage and current waveforms, ϕ is the phase angle between the voltage and current waveforms, and n is the number of ac periods. This yields for any integer $n = (1, 2, 3, \ldots)$

$$P = \frac{V_{peak} I_{peak}}{2} \cos \phi \tag{4.4}$$

As the integration time increases as a function of ac periods, n has a smaller effect on the final result. For example, if 100.5 periods were accidentally integrated (i.e., noninteger $n = 100.5$) instead of 100 or 101 periods, there is only a 0.5% error. If instead $n = 1,000.5$, then the error is only 0.05%. This result for large n is the basis of power measurements. As long as the test system uses many cycles of n to compute the power, it really does not matter if an exact integer number of cycles are used, because the error itself becomes unmeasurable.

Knowing this result, it is easy to predesign the absolute power-level accuracy of a test system measuring a band-limited signal with a bandwidth equal to BW. If the maximum frequency (f_{max}) and the sample rate (f_s) are both known, and the number of sample points is also known, then the worst-case integration period can be calculated.

4.10 Average Power

From Section 4.9, we learned the term *power* means using an averaging time that is "many periods of the highest frequency." The term *average power* means that the energy transfer rate is to be averaged over many periods of the lowest frequency involved. For a CW signal, the highest frequency and the lowest frequency are the same, so the average power and power are the same. However, for an amplitude modulated (AM) wave, the power must be averaged over many periods of the modulation component of the signal.

This can be represented mathematically as

$$P_{AVG} = \frac{1}{nT_l} \int_0^{nT_l} V(t) \times I(t) dt \qquad (4.5)$$

where T_l is the period of the lowest frequency component of $V(t)$ and $I(t)$.

4.11 Pulse Power

When referring to pulse power, the power is averaged over the pulse width, τ. The pulse width, τ, is usually taken to be the 50% rise-time/fall-time amplitude points. The equation for pulse power is give by

$$P_{pulse} = \frac{1}{\tau} \int_0^{\tau} V(t) \times I(t) dt \qquad (4.6)$$

Pulse power averages out such things as overshoot and ringing, which is why it is called pulse power and not peak power. Using (4.5) and (4.6), the equation for pulse power of rectangular pulses can be rewritten as

$$P_{pulse} = \frac{P_{AVG}}{Duty\ cycle} \qquad (4.7)$$

where the duty cycle is the pulse width times the frequency. This can be used to simplify the pulse power measurement to a simple average power measurement that is then divided by the duty cycle. However, care must be taken that the pulse shape is well behaved and understood.

4.12 Modulated Power

When measuring modulated or pulsed power, there are some additional points to consider. The shape of the signal must be understood. For signal shapes that have high peak-to-average ratios, the term *crest factor* is introduced:

$$\zeta = \frac{peak\ value}{rms\ value} \qquad (4.8)$$

The crest factor may be used for voltage or power. For example, the crest factor for a pure sinusoidal signal would be

$$\zeta_{sine} = \sqrt{2} \qquad (4.9)$$

For a pulsed signal the crest factor would be approximately

$$\zeta_{pulse} \approx \sqrt{\frac{1}{Duty\ cycle}} \qquad (4.10)$$

In general, as the crest factor increases, the energy content of higher-order harmonics of the signal increases. To achieve minimal distortion in transmission and recovery of digital signals (i.e., digital modulation with RF waveforms), the signal can become very complex. Large crest factors can cause interference with adjacent channels, add inband distortion, and as a result increase the bit error rate (see Section 5.22).

Additionally, if more complex pulse shapes are being used [e.g., Gaussian pulse shapes used in GSM and Bluetooth or spread-spectrum techniques in code division multiple access (CDMA) technology] [3], then more complex DSP techniques must be employed to determine the pulse or modulated power, and more complex modulated power measurements like adjacent channel power ratio (ACPR) are required.

4.13 RMS Power

The effective, or rms, value of a sinusoidal signal is another commonly used notation to describe the effective voltage, and it also simplifies power calculations when sinusoidal signals are involved. The basic idea is to determine what constant dc voltage when applied across the terminals of a resistor would provide the same amount of electric energy over T seconds as the sinusoidal voltage would. To determine the answer the energy functions for the dc voltage and sinusoidal voltage must be equivalent. The energy of a dc voltage across a resistor is written simply as

$$W_S = \int_{t_0}^{t_0+T} \frac{V^2}{R} dt \qquad (4.11)$$

If V is some dc voltage, then W_s becomes

$$W_S = \frac{V^2}{R} T \qquad (4.12)$$

The energy of a sinusoidal voltage across the same resistor is written as

$$W_o = \int_{t_0}^{t_0+T} \frac{V_o^2(t)}{R} dt \qquad (4.13)$$

Setting the two equations equivalent to one another yields the effective voltage as

$$V_{eff} = \sqrt{\frac{1}{T} \int_{t_0}^{t_0+T} \frac{V_o^2(t)}{R} dt} \qquad (4.14)$$

We now see where the term *rms* is derived. Finding the effective voltage involves taking the square *root* of the *mean* value of the *square* of the function. Thus, the procedure is described as finding the root-mean-square value of the function.

Now, if we substitute V_{eff} into the power equation, it becomes

$$P = \frac{V_{eff}^2}{R} \quad (4.15)$$

Thus, we can conclude that the average power delivered to the load, R, is simply the effective voltage squared, divided by the load resistance, R [4].

4.14 Gain

Historically, simple RF devices were represented by a block diagram similar to Figure 4.3.

In the figure a simple block diagram of a generic RF amplifier is shown. The RF amplifier is represented as the device under test (DUT) with one input, x, one output, y, and having a gain = G. The input, x, and output, y, are functions of frequency in terms of power. The gain, G, is the transfer function and is dimensionless. With this simple block diagram, the transfer function can be written immediately by inspection as

$$G = \frac{y(f)}{x(f)} \quad (4.16)$$

Or equivalently and more useful in decibels as

$$G|_{dB} = y(f)|_{dB} - x(f)|_{dB} \quad (4.17)$$

Equation (4.17) greatly simplifies the mathematics and is more commonly used. To obtain the output power of any amplifier with gain = G, simply add the gain to the input power. Ideally, the gain G is a constant and has no dependencies. In real applications, G is a function mainly of frequency and should more accurately be written as $G(f)$, but instead is simply referred to as G with the understanding that it is highly frequency dependent. Since power is the more commonly used form, (4.17) can be rewritten with x and y replaced by P_{in} and P_{out} respectively, and with the reference to $G(f)$ replaced simply by G:

$$G = P_{out} - P_{in} \quad (4.18)$$

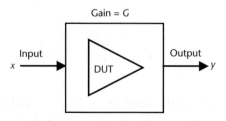

Figure 4.3 Block diagram of an amplifier.

4.14 Gain

Moreover, to determine the output power when dealing with cascaded amplifiers (i.e., two or more amplifiers connected in series), one simply sums each of the gain stages and adds that to input power, P_{in}. Cascading amplifiers is extremely common in RF SOC devices and usually required because it is very difficult to build one amplifier that has both a wide bandwidth and high gain. The total gain of a group of amplifiers in cascade (series) is given as

$$G_{cascade} = G_1 + G_2 + G_3 + \ldots \quad (4.19)$$

where all gain values are in decibels.

Generally speaking, gain is a vector quantity and both the input and output matches of the amplifier must be considered when making a gain measurement, but for the purposes of this discussion, it will be assumed that the mismatches are negligible and that only magnitude values are desired. If the mismatches were not negligible, then S-parameter vector measurements would need to be made to determine the gain. For a more complete academic discussion about gain, refer to Section 4.23.

Given (4.18) and negligible mismatches, power measurements can be used to determine the magnitude gain accurately. A known input power, P_{in}, at frequency f is applied to the DUT, and the output power, P_{out}, at frequency f is measured. The difference is then the gain: $Gain = P_{out} - P_{in}$. Since gain is a function of frequency, a frequency sweep can be performed to obtain gain as a function of frequency.

Figure 4.4 is an example of two amplifiers cascaded together. The first stage gain, $G_1 = 10$ dB, and the second stage gain, $G_2 = 15$ dB. P_{out} is then simply the sum of the gain stages plus the input power P_{in}. If $P_{in} = 0$ dBm, then $P_{out} = 0$ dBm + 10 dB + 15 dB = 25 dBm. However, care must be taken when applying the input signal to ensure that the entire amplifier chain remains in the linear region; otherwise, compression or saturation will distort the final output power. For more information on compression, refer to Section 4.18.

Insertion loss is a measure of the attenuation of passive DUTs. The attenuation may be caused by either the power being converted to heat or by absorption due to an impedance mismatch. In either case insertion loss is similar to gain, only for a passive device, the output power is always less than the input power. The definition of insertion is the reverse of that of gain:

$$IL = P_{in} - P_{out} \quad (4.20)$$

such that insertion loss is always a positive value.

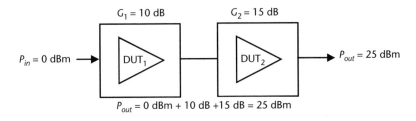

Figure 4.4 Cascaded amplifiers.

When many devices are placed in cascade, (4.19) can be modified to include both active and passive devices (gain and insertion loss):

$$G_{cascade} = G_1 + G_2 + G_3 + \ldots - IL_1 - IL_2 - IL_3 \ldots \quad (4.21)$$

It should be clearly noted that there are two methodologies for performing gain measurements in production testing. From (4.18) it can be seen that both the input and output power levels have to be determined. The output power level is always measured. However, if the test equipment or test system does not have the ability to measure the input power to the DUT, it may be accepted that the requested input power is seen by the DUT. This can potentially speed up the measurement (saving only a few milliseconds, which may be critical in a long test plan); however, one must be aware of the consequences. If the power-level calibration is not accurate or the load board–DUT input impedance matching is poor, then the assumption that the requested power level arrives at the DUT is incorrect. If the load board–DUT input impedance matching is poor, then some of the input power will be reflected back to the source, causing less power to arrive at the DUT.

4.14.1 Gain Measurements of Wireless SOC Devices

Making magnitude gain measurements on RF mixers and SOC devices is more complex because of the higher levels of integration. The complexity comes from the frequency translation that occurs inside the DUT (see Section 2.4 for more about mixers). Figure 4.5 shows a typical block diagram of an RF SOC device. The block diagram depicts the receiver chain, RX, and the transmitter chain, TX. In the traditional RF device, both the input and output frequency were identical, but for the RF SOC device, the output frequency is not the same as the input frequency. Another building block, the mixer, is integrated into the device. The mixer utilizes a local oscillator (LO) to translate the frequency either up or down exactly the same way that test systems do. The LO may be imbedded internally within the RF SOC device, or it may need to be supplied from an external source. When the device is frequency translating, the gain from the input to the output of the DUT is termed *conversion*

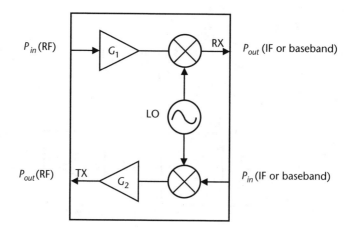

Figure 4.5 RF SOC amplifier.

gain. The transmitted power, P_{out}, and received power, P_{in}, on the antenna's side are at some RF frequency. The transmitted power, P_{in}, and received power, P_{out}, on the processor's side are at some other baseband frequency.

Equation (4.18) still holds, even though the frequency is different at the input and output of the DUT. Rewriting it slightly for the case of a downconverting RF mixer or an SOC receiver yields

$$G_{receiver} = P_{out}\big|_{baseband} - P_{in}\big|_{RF} \qquad (4.22)$$

For an upconverting RF mixer or SOC transmitter the equation becomes

$$G_{transmitter} = P_{out}\big|_{RF} - P_{in}\big|_{baseband} \qquad (4.23)$$

In either case, the complexity is reduced to measuring the input power, P_{in}, at one frequency and measuring the output power, P_{out}, at another frequency.

Complications can arise due to the frequency ranges. The test system may use one method to measure the RF power and another method to measure the baseband power. The increased complexity may require that different hardware be employed to make the necessary measurements. This increases the possible sources of error, as well as complicates the calibration procedure. But, careful and thoughtful system design can negate these extra sources of error.

As an example, consider the traditional two-port RF device discussed previously in Figure 4.3 versus the RF SOC device shown in Figure 4.5. Measuring Figure 4.3 requires only a single downconverting process followed by a digitizer, because both the input and output frequencies are identical. In contrast, measuring Figure 4.5 may additionally require arbitrary waveform generators (AWGs or Arbs) to generate baseband input signals for the TX chain and perhaps a second digitizer to digitize the lower-frequency baseband signals. Instead of a single hardware setup like Figure 4.3, three separate hardware setups (downconverting chain, second digitizer, and Arbs) may be required, and each hardware instrument must be carefully calibrated. The complexity further increases when the input and output matches are not all identical, which is common. It is quite common that the Arbs will need to drive differential non-50-ohm impedances and that the baseband digitizer will need to digitize differential impedances. These additional requirements further increase the complexity of the calibration scheme, making it very difficult to set a desired input power level and measure the correct output power level. As is often the case, external transformers or matching networks are employed on the load board or device interface board (DIB) to match the DUT to the test equipment, and in many instances these external components are unable to be included in the calibration plane.

4.15 Gain Flatness

The concepts of performing power sweeps versus frequency or gain sweeps versus frequency were mentioned above in Section 4.14. We discussed the fact that gain is extremely frequency dependent. So, that raises the question, How much is the gain dependent on frequency. This is usually clearly indicated in the specifications of the wireless SOC device or in the amplifier that is used as a building block in the RF SOC device. Usually, the gain frequency response of an amplifier in an SOC device is desired to be flat with no ripple. Figure 4.6(a–c) shows three ideal amplifiers with ideally flat, ideally downward sloping, and ideally upward sloping gain curves, respectively. The gain curves are of course analog in nature.

In some instances, designers compensate for known nonlinear gain versus frequency behaviors by purposefully slanting the gain response of the device either up or down as shown in Figure 4.6(b, c). This is especially true if the amplifiers are being cascaded with other amplifiers. A downward-sloping gain amplifier can be cascaded with an upward-sloping gain amplifier with the result being a flat gain versus frequency response. If a sloped gain is intentionally designed into the device, then it is often even more imperative to measure the gain slope or gain flatness to ensure the quality of the design.

Figure 4.7 shows a plot of gain response over frequency for an actual device. The gain response across the pass band shows that the gain is not flat, but rather contains some ripple. This rippled gain is more indicative of the behavior of a real RF SOC amplifier or amplifier chain. Examining this figure more closely reveals a maximum gain (G_{max}) point and minimum gain (G_{min}) point. The gain flatness is then simply determined by

$$Gain\ flatness|_{dB} = G_{max} - G_{min} \qquad (4.24)$$

Measuring gain flatness in terms of power can be very simple so long as the following three assumptions hold true:

1. The magnitude gain measurement can be reduced to power measurements.
2. The ripple being measured is large enough to be detected by the test equipment.

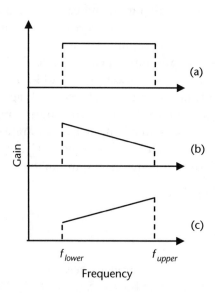

Figure 4.6 Ideal gain slopes: (a) flat gain, (b) downward-sloped gain, and (c) upward-sloped gain.

4.15 Gain Flatness

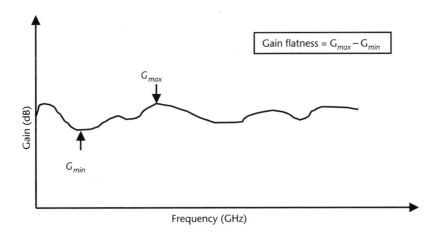

Figure 4.7 Gain flatness measurement.

3. The number of frequency points used to determine the sweep is large enough to provide the required resolution to make an accurate and repeatable measurement.

The gain flatness ripple depends on the bandwidth being examined, but it is usually in the range of tenths of a decibel or less. If the test equipment that is being used has the same or worse uncertainty as the flatness requirement, then it is improbable that the measurement result has any real meaning. In other words, the results might just reflect the uncertainty of the test equipment and not the gain flatness of the DUT at all.

Another point to consider is the frequency spacing used to determine the gain flatness. If the frequency spacing is too large, the G_{max} or G_{min} point could be missed completely. This could result in a lower gain flatness result being measured that would pass the test specification. If that happens, unfortunately a failing device may have just been shipped to the customer. In contrast, if the frequency spacing is extremely small, this issue is completely eliminated. However, smaller frequency spacing adds additional test time. Once again, test time versus measurement accuracy rears its ugly head, and the test/application/product engineer is called in to save the day.

4.15.1 Measuring Gain Flatness

The simplest case would be, if the SOC device requires an RF input and RF output, then continuous wave RF input tones are injected into the SOC device, and the output power is measured.

If the SOC device has a higher level of integration so that mixed-signal type instruments are needed, then perhaps digitizers would be required to capture baseband-type signals, and arbitrary waveform generators would be required to inject baseband-type signals. As a colleague of mine recently said, "The only difference between RF and mixed signal is that RF requires a signal-separating device, mixed signal does not. The types of digital signal processing necessary for both applications are essentially equivalent."[1]

Does that mean that if the three assumptions mentioned in Section 4.15 are not true, an accurate gain flatness cannot be determined. No it does not. If a vector gain measurement is required for accurate results, then it is still possible to make an accurate measurement. It just means that a test setup capable of measuring vector quantities is required.

A final thought on gain flatness measurements: Since test time is one of the most critical elements to be minimized, it makes sense to utilize as much of the test system's available bandwidth when making integrated and sophisticated measurements. Gain flatness is not exactly a sophisticated measurement, but it does provide a good medium to demonstrate this bandwidth-utilization concept.

SOC devices, by definition, are integrated devices (i.e., system on a chip). Then, by definition the test system must also have the full spectrum of test equipment (dc supplies, digital pins, digitizers, arbitrary waveform generators, RF sources, time interval analyzers, and so forth). Assuming that the test system has all of these capabilities, then a simple gain flatness measurement can be made with the following approach:

- An arbitrary waveform generator, either alone or in conjunction with an RF source can be used to inject a flat multitone input signal to the device.
- The output is captured using a wide bandwidth digitizer, and the results are processed to obtain the gain flatness.

Figure 4.8 shows an example of what the input and output signals might look like.

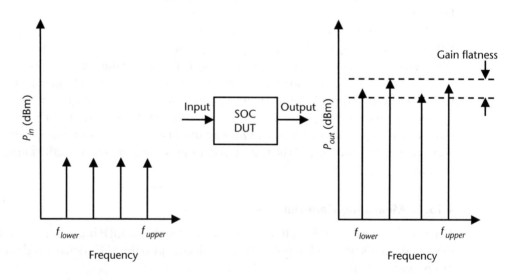

Figure 4.8 Multitone for gain flatness.

---

1. Scott Chesnut, Agilent Technologies, personal communication to author, February, 2003.

4.15 Gain Flatness

Testing the device in this manner eliminates the need for RF frequency or power-level changes, but more importantly, and this is perhaps a little more subtle, the higher bandwidth utilization increases the efficiency of the test equipment or test system. The utilization increase can be directly related to the cost of test of the chip.

4.15.2 Automatic Gain Control Flatness

In Chapter 2, the variable gain amplifier was introduced. One such implementation makes use of automatic gain control (AGC). Figure 4.9 shows two sets of data: the first is for an ideal AGC device; the second is for an actual device. In contrast to the previous discussion, these plots are of power versus time, where at certain instances in time the gain state of the amplifier is changed. In this case, the gain state is stepped from a low value to a higher value; thus for a constant input power level, the output power is as shown.

Even if the majority of the testing is power based, the test time rises on the order of 2^n (n being the number of gain states). Now, instead of making one frequency sweep of gain-versus-time, 2^n frequency sweeps of gain-versus-time are required.

Testing all n gain states would not be cost effective. Luckily, RF design engineers are extremely talented and creative, and most wireless designs can be guaranteed to work even if only the most significant bits (MSBs) of the AGC are tested.

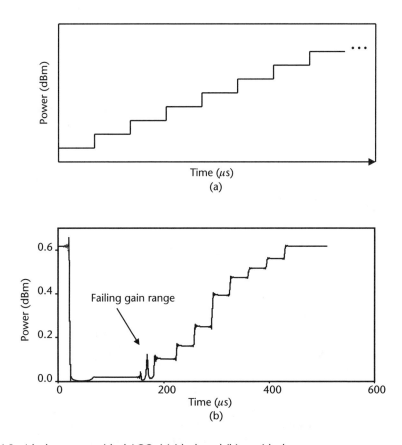

Figure 4.9 Ideal versus nonideal AGC: (a) ideal, and (b) nonideal.

This can still take quite a bit of test time, depending on the number of sample points that are used for each of the sweeps.

An ideal AGC device exhibits the following behavior:

- The gain-level changes from one level to the next are instantaneous.
- There is no overshoot or undershoot directly after the gain-level changes.
- The gain-level changes are linear with the control voltage.

An actual AGC device exhibits the following nonideal behavior:

- Overshoot and undershoot occur during a gain transition;
- Rise and fall times that are not instantaneous;
- Varying gain step sizes.

Figure 4.9(b) is a power-out-versus-time plot of a real wireless SOC AGC device. The plot is more indicative of what a typical wireless AGC measurement sweep might look like. The ringing that occurs during each gain transition is apparent. Also notice near the bottom that one of the bits does not seem to be functioning properly.

There are key parameters that are of interest when examining such a graph:

- DNL/INL (See Chapter 6 for more information on DNL and INL):
  - Gain level. Each gain level should be correct. This is similar to making DNL (differential nonlinearity) and INL (integral nonlinearity) tests that are common for mixed-signal ADCs and DACs.
- Gain settling time:
  - The transition time from one gain state to another. This is critical for today's state-of-the-art wireless formats that operate in the microseconds.
  - The settling time, or "bouncing," of each gain state. Each gain state on Figure 4.9(b) has a slight overshoot and undershoot. It is often desirable to measure this parameter to ensure that the gain settling time is within the specification.

To make power-versus-time, gain-versus-time, or frequency-versus-time transition-type measurements invariably means that the test system must be equipped to make triggered time-domain measurements. Most test systems today have this capability already integrated into the test systems. The triggered time-domain area of SOC production testing is quickly expanding and is often a major battleground between automatic test equipment (ATE) vendors. By its very nature, the time domain requires that the user have access to the time-domain data. This usually means that regardless of the measurement device that is used to capture the data, the time samples must still be transferred to a host computer to complete the data processing, and this increases test time.

An example algorithm is provided for a generic SOC AGC device that is being exercised across its entire gain range at a single frequency point.

Pseudocode

- Set up the input signal to the appropriate frequency and power level.
- Set up the output measurement equipment to receive the output signal when triggered.
- Program the SOC AGC to the first gain level and trigger receiver (the receiver at this point continuously captures the output signal).
  - Cycle the SOC AGC to next gain level.
  - Wait long enough to capture the relevant data.
  - Cycle to the next gain level and repeat until the last gain level has been captured.
- Transfer the time-domain data to the host computer for postprocessing.

Example: Calculate power versus time from the voltage time samples

$$P|_{dBm} = 20 \times \log\left(\sqrt{V_R(i)^2 + V_I(i)^2}\right) + 13 \quad (4.25)$$

Equation (4.25) assumes the test system is referenced to 50 ohms, where $V_R(i)$ and $V_I(i)$ are the real and imaginary voltage time samples of the output waveform and i indicates the time index.

4.16 Power-Added Efficiency

Power-added efficiency (PAE) is another common traditional RF measurement that is strictly confined to amplifiers. Although this is a measure of efficiency, it is similar to a gain measurement. However, it differs from ordinary gain measurements in that it takes all of the power input into the device into consideration. As with all wireless devices, battery life is one of the most critical factors. Users are always concerned about the "talk time" and "standby time" that a new emerging mobile phone is advertising. With the roll out of 3G, embedded cameras are now being offered with many of the new 2.5G and 3G mobile phones. These cameras with faster DSPs can utilize the higher data rates offered by the 3G technology and are advertised to offer still pictures as well as streaming video. The cost is of course to the battery. Faster DSPs drain the battery that much faster. Designing and building a better battery is a constant battle. This leads to an understanding of why chip designers and mobile phone manufacturers are often concerned with PAE. If company A and company B produce essentially the same power amplifier (PA), but company A's PA has a higher PAE, this ultimately means a longer talk time can be offered to the end user.

Power-added efficiency is simply defined as the RF output power in watts divided by the sum of the input RF power in watts and dc input power in watts; it is written as

$$PAE(\%) = \left(\frac{P_{RFOUT}}{P_{RFIN} + P_{dc}}\right) \times 100\% \quad (4.26)$$

In a production-test plan this test is essentially a free measurement in terms of test time. P_{RFOUT} and P_{RFIN} have most likely been measured in one of the test plan steps. P_{dc} is calculated from the dc voltage required to operate the device and the operating current measured for the device when RF power is turned on:

$$P_{dc} = V_{supply} I_{operating} \tag{4.27}$$

As an example, a dual mode AMPS CDMA amplifier that is in CDMA mode and is being supplied with 3.5V draws 577 mA. In CDMA mode at room temperature the amplifier has a gain of 28.5 dB and an output power of 28.5 dBm. From (4.35) the input power is found to be 0 dBm and from (4.26) the PAE is calculated to be 35%. This can be taken as a typical value. Notice that the PAE is not strikingly high, but these are in fact competitive specifications. Using the above information, the mobile phone supplier can then determine the talk time based on his particular biasing. It should be noted that PAEs higher than 35% are easily achievable by operating the amplifier in a high-efficiency mode, but these high-efficiency modes would not offer the linear operation that the wireless system requires.

4.17 Transfer Function for RF Devices

Many RF devices, active and passive, that are used as DUTs in production testing are based upon diodes and transistors. These devices inherently contribute to nonlinear device behavior. In an RF amplifier, this is undesirable. However, in mixers, for example, the nonlinear behavior is intentionally designed into the device. A short derivation of voltage behavior in diode-based RF devices is presented here and will be useful in explanations throughout the rest of the chapter.

The definition of the current through a diode is

$$I = I_S \left(e^{\alpha V_{tot}} - 1 \right) \tag{4.28}$$

where I_s is a constant (saturation) current, α is a constant dependent on temperature and the design of the diode or transistor structure, and V_{tot} is the combined ac and dc voltage across the diode.

If the voltage is generalized to contain both dc and ac components, then

$$V_{tot} = V_0 + V_{in} \tag{4.29}$$

where V_0 is a dc voltage and V_{in} is a small-signal ac voltage. Since V_{in} is a small signal, a Taylor series expansion can be used to rewrite (4.28) as

$$I = I_S \left(e^{\alpha(V_0 + V_{in})} - 1 \right) = I_0 + V_{in} \frac{dI}{dV} + \frac{1}{2} V_{in}^2 \frac{d^2 I}{dV^2} + \ldots \tag{4.30}$$

It is often easier to work in terms of voltages rather than currents since they are simpler to measure. If both sides of (4.30) are multiplied by constant R (resistance, or more appropriately, impedance), then from Ohm's Law, (4.30) becomes

$$V_{out} = \alpha_0 + \alpha_1 V_{in} + \alpha_2 V_{in}^2 + \alpha_3 V_{in}^3 + \ldots \tag{4.31}$$

where $\alpha_0, \alpha_1, \alpha_2, \alpha_3, \ldots$, are constants that have absorbed the coefficient values in (4.30).

The term α_0 is a dc term describing the dc parameters of a diode. An amplifier, when working in the linear region is described by the linear term α_1. The higher-order terms are used to describe either the proper nonlinear behavior of a mixer or the undesirable, nonlinear distortion found in an amplifier.

The following sections will refer to (4.31) to introduce more complex measurements of RF devices; however, keep in mind that most of the measurements are still all based on the basic principle of power measurement. Measurements involving the a_1 term are discussed in Section 4.18. Measurements involving the α_2 and α_3 terms are discussed in greater detail in Section 4.20.2.

4.18 Power Compression

Power compression is another very common RF measurement that is being made on today's wireless SOC devices. While mixers can exhibit conversion compression, the measurement is most often made on amplifiers; hence, the term gain compression is sometimes used. Wireless devices must operate over a wide dynamic range. The upper bound of the dynamic range is often specified with the 1-dB compression point. An example of this is the maximum gain BER specification of Bluetooth. In that case, the Bluetooth device is being overdriven, causing the receiver to go into compression, and then the BER measurement is made.

Generally speaking, any compression point could be specified (i.e., 2-dB compression or 3 dB-compression), but the 1-dB compression point is most commonly used. It should be noted that the choice of specifying a 1-dB compression point is arbitrary, and device specifications could call out a 2-dB or 3-dB compression spec. There is not necessarily an equivalence or direct correspondence between these numbers. For example, knowing the 1-dB compression point does not imply knowing the 2-dB or 3-dB compression point.

Based on a power series expansion, the voltage transfer function of any amplifier is nonlinear and is written as [5]

$$V_{out} = \alpha_0 + \alpha_1 V_{in} + \alpha_2 V_{in}^2 + \alpha_3 V_{in}^3 + \ldots \tag{4.32}$$

where $\alpha_0, \alpha_1, \alpha_2, \alpha_3, \ldots$, are coefficients of the amplifier. These coefficients are what the designers are attempting to optimize for their specific application. The α_0 term is the dc level output and is usually easily filtered out if the wireless SOC device uses a superheterodyne structure. If the structure is zero-IF (ZIF), the α_0 term is much more difficult to filter. However, techniques have been developed to cope with this problem. The α_1 term is the small-signal gain of the amplifier. In many cases it is usually just referred to as gain. The other a terms are, for the most part, designed to be as small and negligible as possible. [Note: It is interesting to point out that if the designer is designing a frequency translating device (a frequency doubler for example), then the designer is attempting to obtain the largest α_2 term, while minimizing the other terms.]

However, the α_2 and α_3 terms can still be significant, and these terms introduce nonlinearities into the wireless SOC design. Measurements involving the α_2 and α_3 terms are discussed in greater detail in Section 4.20. For the moment, let's assume that the application is a high-gain amplifier and that the α_0, α_2, α_3, and above terms are negligible, so that the transfer equation reduces to

$$V_{out} = \alpha_1 V_{in} \tag{4.33}$$

where α_1 is the gain of the SOC amplifier.

If the examination of the device is limited to a single frequency (i.e., one single continuous wave tone), the output voltage continuous wave is simply the input voltage wave multiplied by α_1. Ideally, the input voltage can be continuously increased, and the output will track linearly always being multiplied by the gain (a_1). However, at some unknown input point, the output will no longer track linearly, because the device will slowly go into compression. Try turning up the volume on your stereo system. At some point the speaker (amplifier) will not be able to go any louder, and you will start to hear noise. It is desired to measure this compression point. Different amplifiers have different compression points. Low noise amplifiers (LNAs), for example, are designed to have a large gain and to amplify extremely small signals. The caveat is, however, that they cannot amplify an already large input signal. If the LNA receives an input signal that is too large, it will go into compression and distort the signal, and all of the data could be lost.

Figure 4.10 is a block diagram of a generic amplifier that is in compression.

The output signal is being distorted and is shown as clipping in the time-domain graph.

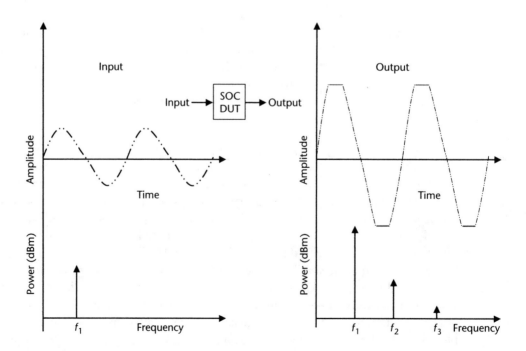

Figure 4.10 Clipping of an SOC amplifier.

4.18 Power Compression

The extra energy, instead of being amplified by the a_1 gain term, goes into the harmonics of the output signal.

How can power compression be measured? As with any SOC measurement, power compression can be made several different ways, but the simplest approach is often the best approach. A wireless SOC device will usually have an input compression specification. If the 1-dB power compression specification is input referred, that specifies the input power level at which the device is 1-dB compressed. If the 1-dB power compression specification is output referred, that specifies the output power when the device is 1-dB compressed.

The equation describing the gain at the 1-dB compression point is

$$G_{1\,dB}\big|_{dB} = G_0\big|_{dB} - 1\,dB \tag{4.34}$$

where G_0 is the small-signal gain, or just gain, as it has been termed throughout this chapter. Using (4.34), the output power out can be rewritten in terms of the compression as

$$P_{1\,dB(output)} - P_{1\,dB(input)} = G_{1\,dB} = (G_0 - 1\,dB) \tag{4.35}$$

Given (4.35), the 1-dB compression point can be found by measuring the difference in the output power minus the input power. When that difference is 1 dB less than the linear gain, the 1-dB compression point has been determined.

Figure 4.11 shows a power-in-versus-power-out plot of an ideal-versus-nonideal amplifier.

The dotted line represents the ideal case and shows the gain always being linear. The solid line represents the nonideal (or real) case and starts to fall off at a certain input power level. The circles are fictitious possible measurement points.

A specific algorithm could be:

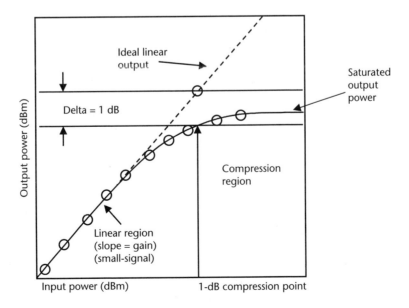

Figure 4.11 Power compression graph.

1. Measure the gain at a power level where the DUT is linear.
2. Extrapolate the linear behavior to a much higher power level.
3. Increase the power in steps, measure the gain, and compare it to the extrapolation.
4. When the difference between 2 and 3 is 1 dB, the measurement is complete.
5. Report either the input or output power level at the point in step 4.

To show that the input-referred and output-referred compression points are related, consider a DUT with nominal small-signal gain of 28 dB that has had the input-referred $P_{1\,dB}$ point determined to be -19 dBm using the above five steps. Rearranging (4.35), $P_{1\,dB(out)} = P_{1\,dB(input)} + G_{1\,dB} = -19 + 28 - 1 = 8$ dBm.

Again, test time must be considered. For some vendors, the exact compression point must be determined. In that case, a binary search usually works best. If a binary search is used, care must be taken not to make the search steps smaller than the accuracy of the measurement equipment. Otherwise, the search algorithm may bounce back and forth between two points, never finding the compression point. For other vendors, test time is more important. For faster test times, the input power is quickly swept, while measuring the output power. A linear interpolation is then used and is sufficient to determine the 1-dB compression point. For Figure 4.11, the circles represent the measurement points and the actual 1-dB compression point is determined by linear interpolation. Care must also be taken when using linear power sweeps. If the device-to-device compression behavior is unpredictable or has a wide standard deviation, then the power sweep might start at too high of an input power level, and the device might already be in compression. Conversely, if the power sweep starts at too low an input power level, then the device may never reach compression.

4.19 Mixer Conversion Compression

A mixer, while considered a nonlinear device, has the same behavior. The only difference is that the input and output of the mixer, taken to be RF and IF, respectively (for example), are at different frequencies. The same algorithms apply, using power measurements at the RF and IF ports of the DUT. As RF input power is increased, IF ouput power should linearly increase. However, at some power level, the IF power begins not to increase as much as the RF input power, and eventually the IF power level deviates from its linearly expected value by 1 dB. As stated before, this point is the conversion compression point.

4.20 Harmonic and Intermodulation Distortion

All devices, whether RF or otherwise, exhibit nonlinear behavior. At times this is part of proper operation, as in the case of an RF mixer. At other times, nonlinear behavior is undesired and a problem that deteriorates the intended performance of a DUT. There are two principle types of distortion in RF and SOC devices for wireless communications. Those two types of nonlinear behavior, or more appropriately distortion, are harmonic distortion and intermodulation distortion. The definition of harmonic distortion is relatively simple, and its calculation is straightforward. Intermodulation distortion requires a slightly more theoretical definition to provide a full understanding.

4.20.1 Harmonic Distortion

Harmonic distortion occurs when some of a DUT's intended power is unintentionally transferred from a desired frequency to a higher frequency multiple of the fundamental frequency. This typically happens at higher power levels.

Harmonic distortion is defined or tested by the application of a single-tone (frequency) sinusoidal waveform. Consider what happens if the input voltage waveform to a DUT is a single-tone frequency, $V_{in} = \cos(\omega t)$, where ω can be any arbitrary frequency. From (4.31), V_{out} is then described by

$$V_{out} = \alpha_0 + \alpha_1 \cos(\omega t) + \alpha_2 \cos^2(\omega t) + \alpha_3 \cos^3(\omega t) + \ldots \quad (4.36)$$

Applying trigonometric identities would show that each term (second term, third term, fourth term, …) can be rewritten as a multiple of the fundamental frequency ω. For example, the second-order term can be rewritten as

$$\alpha_2 \cos^2(\omega t) = \alpha_2 \left(\frac{1 + \cos(2\omega t)}{2} \right) \quad (4.37)$$

These higher-order terms are called harmonics and are classified by their order. The order is an integer and is taken to be m.

Thus for any $V_{in} = \cos(\omega t)$, the output will consist of all harmonics, $m\omega$, where m is an integer going from minus infinity to infinity. In real applications, only the first few harmonics are of any concern and are measured. However, even low-order harmonics can quickly become impossible to measure for some common wireless SOC devices if the test equipment is incapable of measuring very high frequencies. For example, a WLAN 802.11a device has a fundamental operating frequency in the range of 5.6 GHz [2]. The second and third harmonics are already 11.2 GHz and 16.8 GHz respectively. Many of today's automated test equipment manufacturers have great difficulty in designing and building a test system that is capable of making these second- and third-order harmonic measurements. These harmonics are generally, but not always, outside the pass band of the SOC device and can often be filtered.

All of the higher-order terms can be written in terms of the fundamental frequency, and from that it is immediately noticed that each higher-order term is really the fundamental frequency (ω) multiplied by the order (e.g., 2 ωt for the second order in this case) of the term.

Harmonic distortion is specified (and tested) at a specified output power of the DUT. For example, if all of the desired power coming from a DUT were contained in a single tone at the fundamental frequency when the device was operating at low power levels, then when the device power level was increased, if nonlinearities came into play, the power would begin to be seen at the second, third, and so forth, harmonics, taking away from the power intended to be at the fundamental frequency.

A measure of harmonic distortion is total harmonic distortion (THD). It is the relative power contained in all harmonics of a signal expressed as a percentage of the fundamental signal power. It is a measure of how well the device converts energy to the desired fundamental signal versus the undesired harmonic signals. It can be defined as follows:

$$THD(\%) = \sqrt{\frac{D}{S}} \times 100\% = \frac{\sqrt{(V_{harmonic2})^2 + (V_{harmonic3})^2 + (V_{harmonic4})^2 + \ldots}}{V_{fundamental}} \times 100\% \quad (4.38)$$

where $V_{harmonic2}$, $V_{harmonic3}$, $V_{harmonic4}$, ... are the voltage amplitudes of the second, third, fourth, ... harmonics respectively. $V_{fundamental}$ is the voltage amplitude of the desired fundamental signal. D is the total distortion power in watts, and S is the desired signal power in watts.

Total harmonic distortion, THD, is also commonly written in terms of dBc as

$$THD(dBc) = 10 \log_{10} \frac{D}{S} = 20 \log_{10} \sqrt{\frac{D}{S}} \quad (4.39)$$

The "20" log() multiplier comes from the fact that RF engineers always speak in terms of power. THD(%) is defined as the square root of the total distortion power divided by the signal power, and this needs to be written in terms of power before converting to a dB form.

Notice that the desired result is to have a THD(%) that is equal to 0%. A 0% THD equates to the numerator being equal to 0 and to having a perfect device with no distortion. THD is usually measured by making several simple power measurements because the fundamental frequency order is usually in the gigahertz region and the harmonics are too far apart for use of a wide bandwidth digitizer.

Signal, noise, and distortion (SINAD) is a measure of the quality of a received signal and is really just another variation of total harmonic distortion. The definition of SINAD in decibels is

$$SINAD(dB) = 10 \log_{10}\left(\frac{S+N+D}{N+D}\right) \quad (4.40)$$

where S is the signal power (watts), D is the distortion power (watts), and N is the noise power (watts).

Ideally, the distortion and noise powers would be zero. For zero noise and zero distortion (or noise and distortion that approach zero), the SINAD(dB) equation would reduce to

$$SINAD(dB) = 10 \log_{10}\left(\frac{S+0+0}{small+small}\right) = 10 \log_{10}(very\ big\ number) \quad (4.41)$$

and the end result would be a large number that would indicate that the device converts energy very efficiently, having almost zero distortion and adding almost zero noise.

If the distortion of one device versus a second device is higher, then the overall SINAD result will be lower, indicating that the first device is not as efficient. This happens because the distortion is both added to the numerator, but then divided by the denominator. The same thing happens for the noise.

As an example, let $S = 1$ and consider that there is zero noise and that the distortion power is 1/10th of the signal power; then

$$SINAD\,(\text{dB}) = 10\log_{10}\left(\frac{S+0+0.1S}{0+0.1S}\right) = 10\log_{10}(11) = 10.4 \qquad (4.42)$$

Now, doubling the distortion to 1/5th of S yields

$$SINAD\,(\text{dB}) = 10\log_{10}\left(\frac{S+0+0.2S}{0+0.2S}\right) = 10\log_{10}(6) = 7.78 \qquad (4.43)$$

Equation (4.64) yields a SINAD number that is smaller than (4.42) by 2.6 dB. This gives a good indication that the distortion plus noise power has increased by approximately two times or that the fundamental power has decreased by two times. In any case, the efficiency has been reduced in terms of power by a factor of two.

SINAD is often employed for baseband measurements while total harmonic distortion is often employed for RF measurements, but with today's wireless SOC devices this RF-versus-baseband distinction is continually growing less important. The SOC engineers of tomorrow will have a stronger grasp of the fundamentals and be able to apply them across multiple testing disciplines.

4.20.2 Intermodulation Distortion

Intermodulation has many names, including third-order intermodulation product (IP3), third-order intermodulation (IM3), and just plain intermodulation product (IP). In any case, IP3, IM3, or IP all refer to the third-order term $(a_3 V_{in}^3)$ that is described in (4.31) and (4.32). Another common term is the third-order intercept (TOI), which is given by a specific value. TOI is a parameter that is directly related to intermodulation measurements; it is the explicit power-level value that specifies the third-order intermodulation power intercept point (IP3). Recall that the values for IP3 or IM3 are stated in units of dBc, or decibels below the carrier power level, and are a relative power measurement. TOI values are in dBm and are either the input power to the DUT (for receivers) at the IP3 point or the output power of the DUT (for transmitters) at the IP3 point.

The single-tone description of the previous section pertains to the topic of harmonic distortion and only reveals part of the wireless SOC picture. Modern wireless SOC devices use multiple tones and multiple modulation formats to squeeze as much information as possible into the channel bandwidth. Let's take a look at a more complicated input waveform, say a two-tone signal $V_{in} = (\cos\omega_1 t + \cos\omega_2 t)$, where ω_1 and ω_2 are closely spaced arbitrary frequencies. Equation (4.31) becomes much more complicated and is rewritten as

$$V_{out} = \alpha_0 + \alpha_1\{\cos(\omega_1 t) + \cos(\omega_2 t)\} + \alpha_2\{\cos(\omega_1 t) + \cos(\omega_2 t)\}^2 +$$
$$\alpha_3\{\cos(\omega_1 t) + \cos(\omega_2 t)\}^3 \ldots$$

$$(4.44)$$

Again using trigonometric identities, following the format of (4.39), the output can be written in terms of all harmonics of the form $mw_1 + nw_2$, where both m and n are positive and negative integers. The orders can then be defined by

$$order = |m| + |n| \qquad (4.45)$$

The second-order term, or V_{in}^2, will create harmonics at the following frequencies: $2\omega_1$, $2\omega_2$, $\omega_1 - \omega_2$, and $\omega_1 + \omega_2$. (Note: There are four combinations of m and n that when added give the order value of two. For the third-order term there are six terms, and so forth.) All of these frequencies are far away from the fundamental frequencies w_1 and w_2, so it is usually easy to filter them out.

The third-order term, or V_{in}^3, will create harmonics at the following frequencies: $3\omega_1$, $3\omega_2$, $2\omega_1 + \omega_2$, $2\omega_2 + \omega_3$, $2\omega_1 - \omega_2$, and $2\omega_2 - \omega_1$ (notice the six terms). The first four terms are again relatively far away from the fundamental frequencies ω_1 and ω_2, so they can easily be filtered.

The last two terms ($2\omega_1 - \omega_2$ and $2\omega_2 - \omega_1$), however, are very close to ω_1 and ω_2. These two terms cannot be filtered and are commonly referred to as the third-order intermodulation products, or simply as intermodulation products. These two terms are the terms that are specified by the IP3 specification of the RF or wireless SOC device.

The even-ordered terms follow the second-order pattern, while the odd-order terms follow the third-order pattern. Thus, even-ordered terms above the second are almost always ignored because they can be filtered. Rarely, but sometimes, the fifth- or seventh-order terms must also be considered, but the dominant contributions come from the lowest-order terms.

How are these two terms measured? A single point is used to define the IP3 point, or intermodulation point. Figure 4.12 is an intermodulation intercept graph showing output power versus input power of an arbitrary wireless SOC device. The gain (small signal), IP2 (second-order intermodulation product), and IP3 (third-order intermodulation product) are all shown on the graph.

A power-out-versus-power-in plot of small-signal gain has a slope of one. This is shown on the plot. The second-order term has a slope of two and the third-order term has a slope of three. These slopes can be determined by direct inspection of the equations. Notice that it is impossible to measure either the IP2 or the IP3 point directly, because the device would enter into compression long before you could physically measure either point. As the input power is continually increased, the device will enter into compression. The output waveform will start clipping, and the extra energy will be diverted into the higher-order harmonics. The linear gain curve must be extended to find the crossing point of the second- and third-order products. Notice that the IP3 point intercepts the linear curve before the IP2 point. The graph highlights that a high IP3 number is desired. The higher the IP3 number, the less distortion the device exhibits.

4.20 Harmonic and Intermodulation Distortion

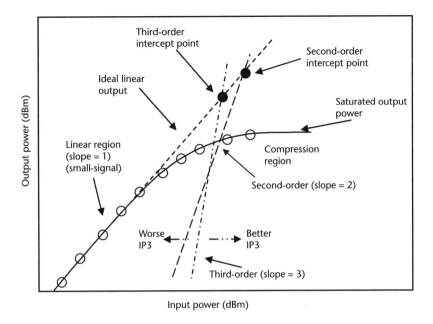

Figure 4.12 Intermodulation intercept graph.

Another common way to display intermodulation products is to use a power-out-versus-frequency plot. A plot of power out versus frequency is shown in Figure 4.13 and is often used when first setting up the measurement to ensure that the test equipment is not measuring the noise floor of the system.

In the figure both fundamental power tones and both third-order power products are shown. Additionally, the noise floor is indicated. The noise floor of a receiver defines the absolute minimum power level that can be accurately measured.

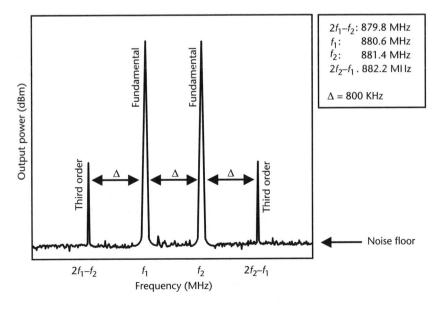

Figure 4.13 Intermodulation plot (power out versus frequency).

Any signal levels that are below the noise floor of the system are completely masked by the noise floor and essentially nondetectable. Think of being at a sporting event where the noise from the fans is so loud that you cannot hear the person next to you talking. That is an example where the noise floor is too high and is completely masking the person's conversation, making it nondetectable.

In the example in Figure 4.13, the third-order products are well above the noise floor of the receiver and are easily measured. However, what happens if we continually reduce the input power to the device in steps of 1 dB? What happens to the fundamental power and the third-order product power? Recall Figure 4.12 where the slope of the fundamental and slope of the third order are shown to be one and three, respectively. This means that with a 1-dB reduction of the input power, the fundamental tone will reduce by 1 dB, whereas the third-order product power will reduce by 3 dB (the converse is also true). If the input power is continually reduced in 1-dB steps, the third-order product power will continually reduce in steps of 3 dB, and eventually the third-order power level will fall below the noise floor of the receiver, thereby becoming nondetectable and invalidating the measurement results. This is a common mistake that is made by engineers new to the RF or wireless field. To correct this, the engineer has three options.

1. Increase the input power until the third-order product power is well above the noise floor (not always possible due to device specifications).
2. Increase the averaging used in the measurement if the third-order power is slightly above or below the noise floor. This has the effect of reducing the noise floor, making the third-order tone detectable.
3. Decrease the measurement bandwidth being used to measure the third-order power. This also has the effect of reducing the noise floor.

The IP3 point is then given by

$$IP3 = P_{fundamental} + \frac{P_{3rd}}{2} \qquad (4.46)$$

where $P_{fundamental}$ and P_{3rd} are the powers of the fundamental tone and third-order tone, respectively. These two points are shown on the graph. There is some ambiguity in (4.46) because the equation does not specify which third-order powers to use. Theoretically, the two fundamental input powers will always equal one another, and the two third-order products will also always equal one another. In general, this assumption is usually valid, so either tone power may be taken. However, in some cases the test equipment may exhibit a calibration problem, or it may have an uncharacterized filter behavior that causes the two fundamental input powers not to be equal at the input to the SOC device. For such cases, an investigation must be performed. If the cause is calibration related, then perhaps an input power correction applied to one of the input tones will achieve equal power input tones.

Lastly, some SOC devices will not have internal flat filter responses, or they may exhibit frequency-dependent nonlinear behavior so that the third-order tones will not be equal. In such cases, it is a simple matter to measure both of the third-order tone output powers and both of the fundamental powers and use the worst-case power; this does, of course, increase the test time for this measurement, because additional power measurements must be made.

4.20.3 Receiver Architecture Considerations for Intermodulation Products

Narrowband systems such as Global Systems for Mobile Communications (GSM) largely employ superheterodyne radio architectures. Superheterodyne radios have a dual frequency downconverting process, which includes multiple filter stages. These multiple filter stages allow the dc term and second-order (IP2) term to easily be filtered out. Thus, as explained above, the third-order (IP3) term, which cannot be filtered out, is a critical factor and often a required measurement.

However, for the Universal Mobile Telephone System (UMTS), which is a wider-bandwidth system employing spread spectrum, it is highly probable that many zero-IF radio structures will be utilized. The main advantage of the zero-IF design over the superheterodyne design is that zero-IF has a cheaper bill of material (BOM) because no expensive off-chip second-stage IF filtering is needed. With the wider bandwidth of UMTS, it is possible to use ac couplings in the receiver to reduce the dc offsets and some of the IP2 products. The downside is that the IP2 products will still be the dominating factor and likely a required measurement item of the test list [6].

4.21 Adjacent Channel Power Ratio

The adjacent channel power ratio (ACPR) is defined as the ratio of the average power in the adjacent frequency channel to the average power in the transmitted frequency channel. The adjacent channel leakage ratio (ACLR) is synonymous with ACPR and often the two terms are used interchangeably. Depending on the context, the acronym ACPR has been taken to mean either adjacent channel power ratio or adjacent channel protection ratio. To resolve this ambiguity, the Third Generation Partnership Project (3GPP), which covers all GSM [including Generalized Packet Radio Service (GPRS), Enhanced Data for GSM Evolution (EDGE), and W-CDMA specifications] has introduced three new terms: ACLR, adjacent channel selectivity (ACS), and adjacent channel interference ratio (ACIR). ACLR is a measure of transmitter performance. It is defined as the ratio of the transmitted power to the power measured after a receiver filter in the adjacent RF channel. Both the transmitted power and the received power are measured with a filter response that is nominally rectangular with a noise power bandwidth equal to the chip rate. This is what was formerly called adjacent channel power ratio.

ACS is a measure of receiver performance. It is defined as the ratio of the receiver filter attenuation on the assigned channel frequency to the receiver filter attenuation on the adjacent channel frequency [7]. This text will use ACPR. ACPR is an important figure-of-merit measurement item that is utilized in CDMA technology because of the complex signal structure of CDMA.

4.21.1 The Basics of CDMA

Code division multiple access (CDMA) is the adopted 3G technology for the next generation of mobile communications devices. CDMA differs considerably from other discrete frequency analog systems like GSM that employ time division (TD) and frequency division (FD). Traditional GSM technology has two disadvantages. The first disadvantage is a limited frequency reuse plan. Each mobile user in a cell is allocated a specific frequency and is allowed to transmit only on that specific frequency. Another user in an adjacent cell is not allowed to reuse the same channel frequency, even though he is located in a different cell. The reason for this is that both users could be near each other (at the cell border separating the two cells, for example), and they would interfere with each other. Therefore, the frequency reuse plan is limited, and this impacts the provider's capacity. The second limitation is that the limited frequency reuse plan makes roaming difficult, and calls are often dropped (the signal connection is lost). As the user crosses from his current cell to the adjacent cell, his mobile is forced to change its channel frequency, and this can interrupt the connection.

CDMA, in contrast, completely eliminates this conundrum, thereby increasing the overall cell capacity by better than a factor of four without increasing the infrastructure cost. Instead of discretely allocating frequencies, in CDMA a spread-spectrum technique is used that enables adjacent cells to share the entire frequency allocation. CDMA assigns a unique pseudorandom noise signal to each user. The rate of the pseudonoise signal is called the chip rate, and it is much higher than the actual data rate. The desired signal (user's voice or data) is multiplied by the pseudonoise signal, which spreads the information across the bandwidth of the pseudonoise signal, before it is transmitted. Figure 4.14 shows a graph of the desired data signal, the pseudorandom noise signal, and the multiplication of the two signals [8, 9].

In this example, notice that the chip rate is 10 times faster than the actual data rate. The receiver multiplies and integrates the received signal by the same pseudonoise signal to recover the desired signal. As long as the transmitter and receiver are synchronized and both are using the same pseudonoise signal, the desired signal can be recovered. Additionally, transmitters and receivers will not interfere with each other because each pair is assigned a unique noise signal, and the noise signals are orthogonal to one another. The noise signal acts as a coding mechanism and is often referred to as a coding sequence instead of a pseudonoise sequence.

It should now be apparent why an ACPR measurement is required for devices employing spread-spectrum technology. It is critical that the spectrum spreading be

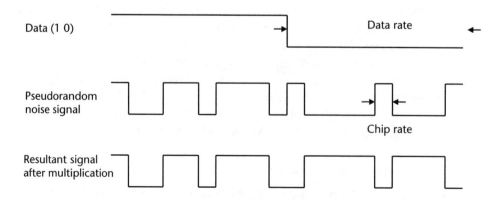

Figure 4.14 CDMA chip rate.

precisely controlled so as not to interfere with other devices and other technologies. ACPR is a good figure of merit that measures the spreading quality of a system.

4.21.2 Measuring ACPR

Figure 4.15 is an example plot of a modulated power out versus frequency for a typical 3G device.

This figure is the result of making an ACPR measurement. The measurement result is realized by downconverting the modulated RF output signal to an IF signal, digitizing the complex time voltage values, and finally performing a fast Fourier transform (FFT) on the complex time array to obtain the final result. If a finer frequency resolution is needed, then a higher number of FFT points can be computed. If more accuracy is required, then a higher number of time samples can be captured for processing. However, increased accuracy versus increased test time is a constant battle regarding production test. The test/application/product engineer must apply all of his skills and knowledge to determine the best trade-off of accuracy versus test time.

In simpler terms, ACPR is a modulated power-out measurement. The plot suggests that the test system must have a wide-enough bandwidth available (for 3G the channel bandwidth is 5 MHz) to make the measurement. If the test system has both a digitizer with a wide-enough bandwidth and a large-enough dynamic range, then the measurement can be performed with a single digitizing capture. This is highly desirable, because it dramatically reduces test time without trading accuracy [3].

If the test system does not have a wide bandwidth digitizer, or there is not enough dynamic range to encompass the range of channel powers, then multiple digitizing captures must be performed. However, this is not desirable because of the

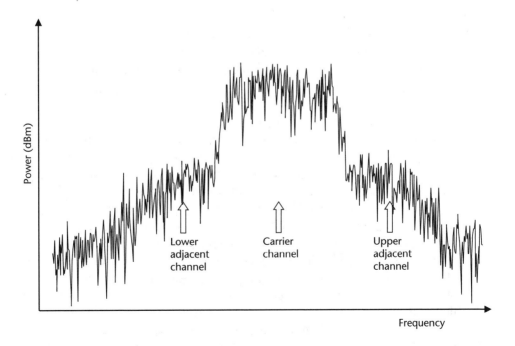

Figure 4.15 Adjacent channel power plot.

disadvantage of additional test time required to make the measurement. As an example, refer to Figure 4.15 and notice three channel powers are shown: lower adjacent channel, carrier channel, and upper adjacent channel. Now, consider a system that does not have a wide bandwidth digitizer (greater than 15-MHz bandwidth), such that multiple captures (in this case, three separate captures) must be made to obtain the carrier power and the lower and upper adjacent channel powers. Some of the incurred extra test time comes from the overhead of the digitizer's having to communicate with the host computer three times, but the largest contributor to increased test time comes from the test system's LO (RF source) having to tune to three different channel frequencies. Changing the frequency or power level of an RF source is generally one of the more time-consuming procedures in making RF measurements, especially if the measurement requires an RF source with low phase noise as low-phase-noise sources are typically the slowest. Depending on the type of source that is required to make the measurement, the test time due to the RF source can be greater than 50% of the total time of the ACPR/ACLR measurement. This means that each retuning of the LO source adds at least 50% more test time for that particular measurement. For characterization, this is not really an issue, but for production testing, where there is constant cost pressure to reduce the test time per device by even a few microseconds, the extra test time cannot be justified.

4.22 Filter Testing

If only the roll-off characteristics are desired, then filter testing can be thought of as a simple power-out test, a power-compression test without the compression, or even a modulated power-out test. The approach used to test a filter depends on many factors:

- Is the test being used for characterization or production?
- Is the group velocity of the pass band desired?
- How many points are needed to define the shape of the filter clearly?
- What bandwidth is required to describe the filter accurately?

If the device is still in characterization, then it may be that many of the filter characteristics are still being examined. If that is the case, then often a chirp signal is used to acquire a complete picture of the filter characteristics. The term *chirp signal* is just an old way of saying that the stimulus signal should have the appropriate bandwidth and be linear-time invariant with a low peak to average power. A helpful way of understanding what low peak to average power means is to consider a stimulus signal that consists of multiple continuous wave (CW) tones or a multitone stimulus. Figure 4.16 shows a power-versus-frequency plot of an arbitrary filter characteristic, noted by the dotted line, and four CW tones used to define the roll-off characteristics of the filter. Using this particular test setup, four points along the curve of the filter are determined, and the general shape of the filter can then be determined with a linear interpolation technique.

Examine the time-domain representation of f_1 and f_4 shown in Figure 4.17. This is an amplitude-versus-time plot. To simplify the discussion, only f_1 and f_4 are shown

4.22 Filter Testing

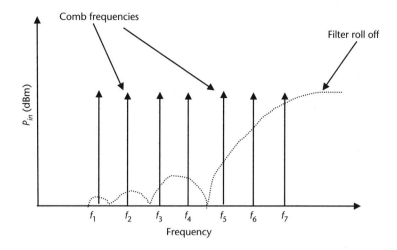

Figure 4.16 Filter testing using comb frequencies.

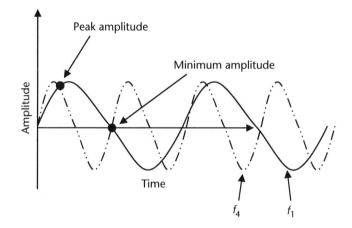

Figure 4.17 Comb frequencies f_1 and f_4 in the time domain.

on the time-domain plot. Additionally, it is assumed that f_4 is twice the frequency of f_1. With f_1 and f_4 drawn on top of one another, it is easy to see that there are distinct points where the sum of the amplitudes is at a maximum (or peak) and other points where the sum of the amplitudes is at a minimum. This is not desirable because it will cause clipping of the summed input waveform and reduce the accuracy of the measurement.

Changing the initial phase of each waveform is necessary to reduce the peak to average power. This can easily be accomplished with any signal-processing tool like Matlab, Mathcad, or even a simple handwritten program.

A multitone stimulus test setup can be created in two ways:

1. Using multiple RF sources tied together with a tone combiner;
2. Using an arbitrary waveform generator capable of the necessary bandwidth required.

RF sources are necessary for testing any of today's wireless SOC devices, but in general RF sources are expensive, so it is usually desirable to minimize the number of RF sources used in a particular test setup. Arbitrary waveform generators are good for generating multitone or modulated signals, and they have the necessary bandwidth that the signals are required to have. However, arbitrary waveform generators are limited in absolute maximum frequency, but can easily modulate other standalone RF sources. So it is common practice to have a test setup with a minimum number of RF sources and a minimum number of arbitrary waveform generators to cover all testing scenarios.

If a test setup does not have an arbitrary waveform generator, then it is still possible to perform a magnitude roll-off test by simply making multiple CW power measurements at the desired frequency points. The limitation is that no phase information can be acquired so that a proper group velocity measurement cannot be made.

4.23 S-Parameters

4.23.1 Introduction

Microwave engineers have used scattering parameters (S-parameters) for many years as a means to characterize devices. They are also used in production testing at times. Often, values that are derived from S-parameter measurements are of more concern in production testing because accurate assessment of S-parameters often requires detailed analysis, and as a result the measurements take more time. Values such as return loss, gain, or isolation are needed. There are other ways to obtain these values than by performing a full S-parameter analysis.

It is very important to have an understanding of what S-parameters are and how they relate to other test parameters. This section will give a definition of S-parameters and demonstrate the relationship of S-parameters to more familiar measured items. Also, some examples will be provided to show how these items are measured in practice.

S-parameters are extensions of transmission line theory where the input and output signals that are used to perform the measurements are waveforms. These waves do not vary in amplitude along a device as would a simple voltage or current. The most important property is that S-parameters contain both magnitude and phase information.

In practice, voltage waves are measured with an instrument capable of measuring vector properties, such as a network analyzer. S-parameter measurements have an advantage over other microwave parameter measurement techniques in that they do not require the implementation of short or open circuits, which can lead to oscillations in devices under test, prohibiting measurement. At high frequencies, the implementation of an open or short circuit is difficult due to the effects of stray capacitances and inductances. Additionally, S-parameter measurements of multiple devices can be cascaded to predict full system performance. Chip designers often use this technique.

4.23.2 How It Is Done

S-parameters are more adequately represented using simple matrix algebra. In general, S-parameters are defined as

4.23 S-Parameters

$$[V^-] = [S][V^+] \tag{4.47}$$

As shown, S-parameters are defined in relation to the incident and reflected voltages. It is important to note, at this point, that most references discuss the "+" and "−" of (4.47) as "incident" and "reflected"; however, the "reflected" wave is often a wave coming out of a port that originated at some other port of the device.

When the matrix notation of (4.47) is expanded, it becomes

$$\begin{bmatrix} V_1^- \\ V_2^- \\ \cdot \\ V_N^- \end{bmatrix} = \begin{bmatrix} S_{11} & S_{12} & \cdot & S_{1N} \\ S_{21} & \cdot & \cdot & \cdot \\ \cdot & \cdot & \cdot & \cdot \\ S_{N1} & \cdot & \cdot & S_{NN} \end{bmatrix} \begin{bmatrix} V_1^+ \\ V_2^+ \\ \cdot \\ V_N^+ \end{bmatrix} \tag{4.48}$$

Any specific S-parameter can be specified as

$$S_{ij} = \left. \frac{V_i^-}{V_j^+} \right|_{V_{k+}=0 \text{ for } k \neq j} \tag{4.49}$$

Equation (4.49) states that S_{ij} is found by driving port j with an incident wave of voltage V_j^+ and measuring the reflected wave amplitude V_i^- coming out of port i [10].

A two-port device has four S-parameters associated with it. In general, an N-port device requires N^2 S-parameters to completely describe it. The typical convention [referring to (4.49)] is that the jth port is where the signal enters the device, and the ith port is where the signal leaves the device. In summary, for a multiport device, S_{ii} is the reflection coefficient of port i when all other ports are terminated in a matched load, and S_{ij} is the transmission coefficient from port j to port i when all other ports are terminated in matched loads.

In practice S_{11} and S_{21} are measured when port 2 is terminated with a perfect characteristic impedance load. S_{12} and S_{22} are measured when port 1 is terminated with a perfect characteristic impedance load. In either case, both the magnitude and phase of the incident and reflected voltage waveforms are measured, and from these the S-parameters are calculated.

4.23.3 S-Parameters of a Two-Port Device

For this discussion, the focus will be on two-port devices such as an amplifier. The standard nomenclature is that port 1 is the input of the device and port 2 is the output of the device. This is the most common type of device in the wireless architecture for which knowledge of the S-parameters may be used in production testing. The following description of two-port S-parameters will reference Figure 4.18.

S-parameter measurements can be classified into two groups, transmitted measurements and reflected measurements. For a two-port device, the four S-parameters are

$$S_{11} = \left. \frac{V_1^-}{V_1^+} \right|_{Z_0 \text{ load on port 2}} \tag{4.50}$$

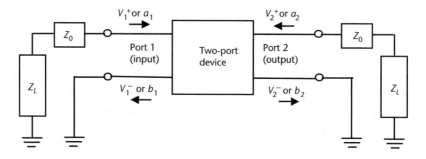

Figure 4.18 Two-port S-parameter device.

$$S_{12} = \left.\frac{V_1^-}{V_2^+}\right|_{Z_0 \text{ load on port 1}} \quad (4.51)$$

$$S_{21} = \left.\frac{V_2^-}{V_1^+}\right|_{Z_0 \text{ load on port 2}} \quad (4.52)$$

$$S_{22} = \left.\frac{V_2^-}{V_2^+}\right|_{Z_0 \text{ load on port 1}} \quad (4.53)$$

In practice, the various S-parameters are found by applying a load of characteristic impedance (often referred to as Z_0) to the various ports of the device and making measurements on other ports. More importantly, the impedance to be used as a termination should match that of the device. This will avoid reflections at the port. Z_0 is typically the same impedance as that of the device and the measuring equipment. In a wireless RF environment, Z_0 is often 50Ω, and in RF cable TV environments, Z_0 is typically 75Ω.

Equation (4.50) states that S_{11} is the ratio of reflected to incident voltages on port 1 when port 2 is terminated by a load of characteristic impedance Z_0. S_{11} is also referred to as the input reflection coefficient.

Equation (4.51) states that S_{12} is the ratio of the voltage coming out of port 1 to the incident voltage on port 2. This is similar (but not equal) to the transfer function of the device, called the reverse transmission coefficient.

Equation (4.52) states that S_{21} is the ratio of the voltage coming out of port 2 (transmitted) to the incident voltage on port 1. This is termed the forward transmission coefficient.

Analogous to S_{11}, S_{22} is the output reflection coefficient, given by the ratio of reflected to incident voltages on port 2.

4.23.4 Scalar Measurements Related to S-Parameters

S-parameters are vector entities derived from vector-based measurements (magnitude and phase). There is, however, a subset of scalar measurements derived from S-parameters. These are shown in Table 4.1.

Table 4.1 Scalar Measurements Derived from S-Parameters

Scalar Reflection Measurements	Scalar Transmission Measurements
Reflection coefficient	Transmission coefficient
Return loss	Insertion loss
VSWR	Gain

The reflection coefficient, Γ, is a scalar reflection measurement that is directly related to S_{11} or S_{22}. From the value measured for the reflection coefficient, other meaningful parameters such as return loss and VSWR can be calculated.

The reflection coefficient is defined as the magnitude of the ratio of reflected to incident voltages:

$$\text{Reflection coefficient } (\Gamma) = \left|\frac{V_{reflected}}{V_{incident}}\right| = \left|\frac{V_1^-}{V_1^+}\right| = |S_{11}| \quad (4.54)$$

or

$$\text{Reflection coefficient } (\Gamma) = \left|\frac{V_{reflected}}{V_{incident}}\right| = \left|\frac{V_2^-}{V_2^+}\right| = |S_{22}| \quad (4.55)$$

Return loss is simply the logarithmic representation of the reflection coefficient:

$$\text{Return loss (dB)} = -20\log(\Gamma) \quad (4.56)$$

Another commonly measured or calculated parameter is the standing wave ratio. It is the ratio of maximum to minimum voltage at a given port. In RF applications it is most commonly referred to as the voltage standing wave ratio (VSWR). It is most simply related to the reflection coefficient as

$$VSWR = \frac{1+\Gamma}{1-\Gamma} \quad (4.57)$$

The transmission coefficient, γ, is defined as the magnitude of the ratio of transmitted to incident voltages:

$$\gamma = \left|\frac{V_{transmitted}}{V_{incident}}\right| = \left|\frac{V_2^-}{V_1^+}\right| = |S_{21}| \quad (4.58)$$

or

$$\gamma = \left|\frac{V_{transmitted}}{V_{incident}}\right| = \left|\frac{V_1^-}{V_2^+}\right| = |S_{12}| \quad (4.59)$$

An alternative definition for insertion loss of a passive device is simply the logarithmic representation of the transmission coefficient:

$$\text{Insertion loss (dB)} = -20\log(\gamma) \quad (4.60)$$

And for an active device,

$$\text{Gain (dB)} = 20 \log(\gamma) \qquad (4.61)$$

An amplifier is a typical two-port device. Each of the four two-port S-parameters has an equivalent meaning when referring to an amplifier operating in the linear region. These are typical key test list items for an RF amplifier. S_{11} is the input return loss. S_{21} is gain from input to output. S_{12} is isolation. S_{22} is the output return loss.

4.23.5 S-Parameters Versus Transfer Function

S_{21} is approximately equal to the transfer function (V_{out}/V_{in}) of a device, but it is not exactly equal. The difference between the two is due to the matching between the device and the equipment making the measurements. Recall that S_{21} is the ratio of the voltage coming out of port 2 to that incident on port 1 when port 2 is terminated with a load of characteristic impedance of the device.

Consider S_{21} in two steps. First, the voltage coming out of port 2 is equal to V_{out} of the transfer function exactly (assuming that the termination on port 2 is that of the device, as the definition suggests). However, the voltage that is incident on port 1 is not always exactly equal to V_{in} of the transfer function. If the input impedance of the device differs from that of the signal source, then the incident voltage to the device will differ from V_{in}, as some of the voltage waveform will be reflected due to mismatch [11].

Example: Two-Port Amplifier

Refer to the generic amplifier in Figure 4.19.

The figure has two ports, so it has only four S-parameters. They are listed below with their common denotations as used in production test with the assumption that all ports are matched:

- S11 (input return loss or input match);
- S22 (output return loss or output match);
- S21 (small signal gain or just gain);
- S12 (reverse transmission or isolation).

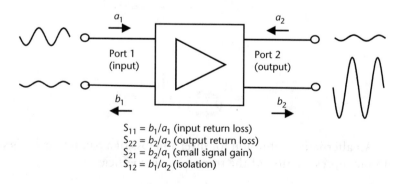

Figure 4.19 S-parameter of generic amplifier.

It is more common to use decibel notation instead of a linear notation. Using decibel notation has the added advantage of providing the reflected power by inspection. For example, if the return loss of an amplifier is 3 dB, then it is immediately known that half of the power is being reflected from the DUT. If the gain of the device were being calculated from the simple gain equation, (output power − input power), as described earlier in this chapter, and the return loss of 3 dB was not known, or the input power was not being properly measured, the gain answer would be completely incorrect. This highlights the danger of oversimplifying vector measurements and ignoring the many assumptions that are made that have provided the much simpler equations that are often used in production testing. This is a very common mistake in production environments and with engineers who are new to RF. In practice, it is best to verify these assumptions when considering an unknown test setup.

It should be noted that a perfectly matched input would have an input reflection coefficient equal to 0. The log of 0 is −infinity. This implies that it is desirable to have a return loss as large as possible. Well-matched two-port wireless communications devices typically have return loss and isolation numbers in the 15- to 25-dB range. As frequencies rise beyond 10 GHz, good return-loss values become much more difficult to achieve. Figure 4.20 is provided to help the reader understand the relationship of all of these scalar S-parameter-based parameters that are often used in production testing.

4.23.6 How to Realize S-Parameter Measurements

Recall that S-parameters are vector quantities. Each parameter has a magnitude and phase associated with it. To further complicate matters, the S-parameters are dependent on both incident and reflected waves. This means that it must be possible to measure both the incident and reflected signals simultaneously. This is made possible through the use of a signal separator. Two common types of signal separators are couplers and bridges. Whether a coupler or bridge is used depends on the application. Each has its advantages and disadvantages.

4.23.7 Characteristics of a Bridge

The characteristics of a bridge are:

- Use to measure reflected signals;
- Higher loss (less power available to the DUT);
- Broadband and effectiveness at lower frequencies.

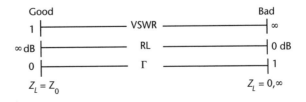

Figure 4.20 Relationship of scalar S-parameter-based measurements.

4.23.8 Characteristics of a Coupler

The characteristics of a coupler are:

- Directional in nature;
- Low loss;
- Good isolation;
- Narrowband and poorer performance at lower frequencies.

What should be apparent from their listed characteristics is that a production-test system needs to have both bridges and couplers available to cover all possible test scenarios. UMTS or 3G fundamental frequencies are in the 1.9-GHz range, whereas WLAN 802.11a fundamental frequencies are in the 5.6-GHz range. Although harmonic measurements are strictly scalar power measurements and are made separately from S-parameter measurements, as was mentioned earlier in this chapter, harmonic measurements are common production-test list items. This means that both scalar (harmonic) and vector (S-parameter) quantities impact the requirements of any production-test system. For second- and third-order WLAN 802.11a harmonics, an upper frequency of 17 GHz is required of the test system. Additionally, depending on the integration level of the DUT, the lower-frequency requirements could be in the tens of megahertz. Directional couplers that have good spectral properties from tens of megahertz to around 18 GHz are not feasible, hence the requirement for both couplers and bridges. A block diagram of a signal separator is shown in Figure 4.21.

A coupler is designed to provide the following ideal properties:

- A small sampling of the incident energy along the incident path;
- A small sampling of the reflected energy along the reflected path;
- Zero energy coupled from the unwanted path.

Real couplers have nonzero energy (unwanted) from the nondesired path that is coupled into the desired path. This leakage (directivity) can be overcome by a vector

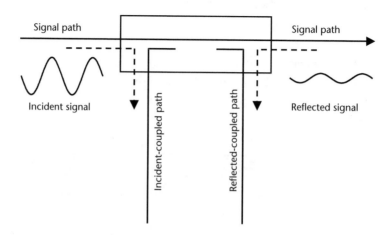

Figure 4.21   Signal separator.

calibration. Once the signal has been coupled, it has physically been separated into its incident and reflected parts. These signals are RF signals, and thus, they are still too high in frequency to digitize directly. Each signal must run through a mixer to be downconverted to a lower frequency. The lower frequency signals can then be digitized directly, but they must be digitized simultaneously not to loose the relative phase relationship between the two signals.

Figure 4.22 shows a block diagram that will physically realize a one-port S-parameter measurement.

Now, to complete all four S-parameter measurements, the realization of another signal separator is needed. This is shown in Figure 4.23.

The figure also represents in the most basic sense a block diagram of today's vector network analyzers.

4.24 Summary

This chapter began with a brief introduction to RF power, along with pioneering developments that were greatly impacted by the two world wars. The importance of having accurate and globally traceable calibrations was stressed. Arguments were then made as to why power measurements are so important before the chapter defined and explained the myriad of power measurements that are performed in production. The logarithmic notations for dB and dBm were presented with examples demonstrating their usefulness. Using power as a basic building block, gain was introduced, and its dependency on frequency was stressed. Measuring gain was applied to various test scenarios using strictly RF devices, as well as SOC devices. The concepts of gain flatness and automatic gain control (AGC) were then explained and applied to SOC devices to highlight the merging of the RF and mixed-signal worlds. Power-added efficiency (PAE) was introduced, and its importance to battery life was demonstrated.

Nonlinear measurements, including power compression, total harmonic distortion, and intermodulation distortion, were discussed, along with various test scenarios covering two of today's SOC devices, the superheterodyne and the zero-IF

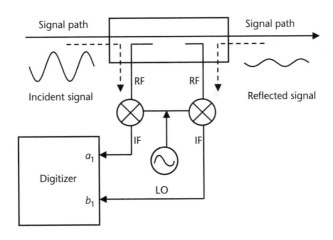

Figure 4.22 One-port S-parameter realization.

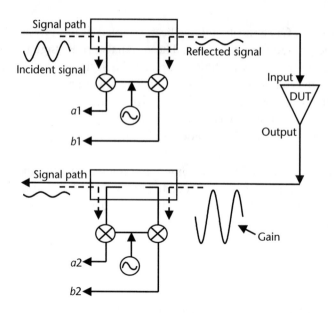

Figure 4.23 Two-port S-parameter realization.

radios. CDMA technology as a 3G technology was briefly discussed and used as an example to introduce and demonstrate the adjacent channel power ratio measurement. SINAD and filter testing were briefly discussed, while similarities between SINAD (in the mixed-signal world) and THD (in the RF world) were pointed out. Finally, a brief description of S-parameter theory was provided. The theory was then applied to a generic two-port device, and explanations and figures on how to physically realize one-port and two-port S-parameter measurements were offered.

References

[1] Agilent Technologies, "Fundamentals of RF and Microwave Power Measurements," Application note 64-1A, 1977.

[2] Van Nee, R., and Prasad, R., *OFDM for Wireless Multimedia Communications.* Norwood, MA: Artech House, 2000.

[3] Agilent Technologies, "Digital Modulation in Communication Systems—An Introduction," Application note 1298, 2001.

[4] Nilsson, J. W., *Electric Circuits.* 3rd ed., Boston: Addison-Wesley, 1990, pp. 373–375.

[5] Smith, J. L., "A Method to Predict the Level of Intermodulation Products in Broadband Power Amplifiers," *Microwave Journal*, Vol. 46, No. 2, February 2003, p. 64.

[6] Minnis, B. J., and Moore, P. A., "Estimating the IP2 Requirement for a Zero-IF UMTS Receiver," *Microwave Engineering*, July 2002, pp. 35–36.

[7] Wolf, J., and Buxton, B., "Measure Adjacent Channel Power with a Spectrum Analyzer," *Microwaves & RF*, January 1997, pp. 55–60.

[8] Nelson, R., "3G Technologies Complicate CDMA Testing," *Test & Measurement World* February 2000, pp. 55–62.

[9] Agilent Technologies, "Performing cdma2000 Measurements Today," Application note 1325, 2000.

[10] Pozar, D. M., *Microwave Engineering* Boston: Addison-Wesley, 1990.

[11]   Witte, R. A., *Spectrum and Network Measurements*, Upper Saddle River, NJ: Prentice Hall, 1993.

Appendix 4A: VSWR, Return Loss, and Reflection Coefficient

Reflection coefficient (Γ), return loss (RL), and voltage standing wave ratio ($VSWR$), are related. The most commonly used parameter in device specifications, however, is VSWR.

Using an S-parameter-based definition, input and output reflection coefficients are defined as [1],

$$\Gamma_{in} = S_{11} + \frac{S_{21} S_{12} \Gamma_{load}}{1 - S_{22} \Gamma_{load}} \tag{4A.1}$$

or

$$\Gamma_{out} = S_{22} + \frac{S_{21} S_{12} \Gamma_{source}}{1 - S_{11} \Gamma_{source}} \tag{4A.2}$$

If the device under test (DUT) is impedance matched to the test equipment then the reflection coefficient of the source and load become zero and the input and output reflection coefficients of the DUT are simply the magnitude of S_{11} or S_{22} (referring to either the input reflection coefficient or output reflection coefficient, respectively):

$$\Gamma_{in} = S_{11} \tag{4A.3}$$

or

$$\Gamma_{out} = S_{22} \tag{4A.4}$$

The relationship between reflection coefficient and return loss is,

$$RL_{dB} = -20 \log_{10}(|\Gamma|) \tag{4A.5}$$

The relationship between VSWR and reflection coefficient is,

$$\Gamma = \frac{VSWR - 1}{VSWR + 1} \tag{4A.6}$$

and it therefore follows that VSWR is related to reflection coefficient by,

$$VSWR = \frac{1 + \Gamma}{1 - \Gamma} \tag{4A.7}$$

VSWR, is often pronounced as a word, "vis-war," rather than the acronym.

Additionally, it is most often written as a ratio, relative to 1, as shown in Table 4A.1, which shows the relationship between VSWR, return loss, and reflection coefficient.

Table 4A.1 Relationship Between VSWR, Return Loss, and Reflection Coefficient

VSWR	RL (dB)	Γ	VSWR	RL (dB)	Γ
1.001:1	66.025	0.0005	1.1:1	26.444	0.0476
1.002:1	60.009	0.0010	1.2:1	20.828	0.0909
1.003:1	56.491	0.0015	1.3:1	17.692	0.1304
1.004:1	53.997	0.0020	1.4:1	15.563	0.1667
1.005:1	52.063	0.0025	1.5:1	13.979	0.2000
1.006:1	50.484	0.0030	1.6:1	12.736	0.2308
1.007:1	49.149	0.0035	1.7:1	11.725	0.2593
1.008:1	47.993	0.0040	1.8:1	10.881	0.2857
1.009:1	46.975	0.0045	1.9:1	10.163	0.3103
1.01:1	46.064	0.0050	2.0:1	9.542	0.3333
1.02:1	40.086	0.0099	3.0:1	6.021	0.5000
1.03:1	36.607	0.0148	4.0:1	4.437	0.6000
1.04:1	34.151	0.0196	5.0:1	3.522	0.6667
1.05:1	32.256	0.0244	10.0:1	1.743	0.8182
1.06:1	30.714	0.0291	20.0:1	0.869	0.9048
1.07:1	29.417	0.0338	50.0:1	0.347	0.9608
1.08:1	28.299	0.0385	100.0:1	0.174	0.9802
1.09:1	27.318	0.0431			

Reference

[1] Agilent Technologies, "S-Parameter Design," Application note 154, 2000.

CHAPTER 5
Production Testing of SOC Devices

5.1 Introduction

This chapter will discuss the many aspects of production testing for wireless SOC devices. Highly integrated wireless SOC devices must supplement the traditional parasitic testing (power, phase noise, S-parameters, ACPR, and so forth) normally accompanying individual building blocks (amplifiers, mixers, filters, and the like) with functional testing of the complete SOC chains (TX, RX). Functional testing of a wireless SOC tests entire chains of the device as a functional block and more closely approximates how the device will be used in the real world. The specific application (e.g., cellular, WLAN, Bluetooth) dictates the functionality that is required of the SOC. As such, it is necessary to have a good understanding of the particular application that the SOC is providing in order to understand the functional testing requirements fully.

For example, a Bluetooth SOC and a WLAN application are two completely different applications targeting different markets. They have vastly varying designs, specifications, and functions. Therefore, while parametric testing of the individual building blocks for cellular, WLAN, or Bluetooth SOC devices can be quite similar, functional testing of a Bluetooth SOC can require vastly different testing versus functionally testing a WLAN or cellular SOC. Thus, it is important to understand the intended function of the SOC.

The various integration levels of SOC devices are described to help the reader break the SOC into the various functional blocks when examining block diagrams of wireless SOC devices. Wireless Bluetooth SOCs are examined in detail, beginning with an introduction to the Bluetooth standard. The origins of frequency hopping and the modulation format of the Bluetooth standard are discussed, along with the various data packets that are supported by the standard and the adaptive power control feature that helps Bluetooth devices combat the congested industrial, scientific, and medical (ISM) band.

The individual blocks that make up a basic radio are described and are in turn used to formulate the major functional blocks of a radio, namely the PLL, TX, and RX sections. The building blocks and testing requirements of a phase-locked loop (PLL) are described since a PLL is a mandatory functional block of any radio. The various test setups that are required for wireless SOC devices and various testing methodologies for PLLs are presented, before we dive into the two key functional blocks, namely TX and RX.

The functional testing requirements for testing the TX and RX functions of wireless Bluetooth devices are explained, and examples are provided along the way to assist in understanding. The major functional tests of a Bluetooth transmitter are explained, and example test setups show how to realize the measurements in a production environment. Additionally, the functional testing of the receiver chain of a Bluetooth SOC is described and bit error rate (BER) is introduced with careful emphasis on the myriad test setups available to realize production BER measurements. Each of the Bluetooth BER measurements is described, and analogies and graphs are presented to assist the reader. Lastly, an introduction to error vector magnitude (EVM) is presented in relation to production testing.

5.2 SOC Integration Levels

A wireless SOC device must have a radio inside of it. This radio will have two essential sections, a transmitter section (TX) and a receiver section (RX). Depending on the integration level of the SOC, the TX and RX sections may share various redundant blocks (for example, the modulator/demodulator or the PLL), and for very high levels of integration, the radio may have multiple radios to cover multiple bands. In any case, it is important to understand the three basic wireless SOC configurations. Table 5.1 contains a list of the three basic configurations found in today's wireless SOC devices.

A brief discussion of each configuration follows.

Case 1: RF-to-RF Configuration

SOC devices with RF-to-RF configurations are the most basic of wireless SOC devices and this configuration would be classified as an RF SOC. An RF-to-RF SOC configuration (or RF-to-IF) is simply a collection of RF building blocks (amplifier, mixer, and filter, for example) manufactured on a single die. The test system and skill set requirements needed to test such a configuration are much more RF based and require little or no knowledge of digital and baseband signal testing. A modulator or amplifier mixer combination is a good example of an RF SOC.

Case 2: RF-to-Analog (Baseband) Configuration

RF-to-analog configurations have a higher level of integration than RF-to-RF configurations. The test system requirements will usually include single or multiple digitizers to capture the analog baseband signals and arbitrary waveform generators to stimulate the DUT with baseband signals. The skill set or knowledge base is also more demanding, because the baseband signal configuration can be either single-ended or differential-ended and can have non-50-ohm impedances, as well as dc offsets. This sort of configuration can be classified as an RF/mixed-signal SOC and are most commonly found in WLAN SOC configurations.

Table 5.1 Wireless SOC Configurations

TX Output Configurations	RX Input Configurations
RF	RF
Baseband (analog differential or single ended)	Baseband (analog differential or single ended)
Digital	Digital

Case 3: RF-to-Digital Configuration

RF-to-digital configurations can be found in many of today's Bluetooth SOC devices. An RF-to-digital configuration includes the baseband decoder that decodes the baseband signal into digital 1s and 0s. To test this configuration's requirements adequately (namely, its bit error rate), the test system must have digital capabilities. The engineer's skill set must include RF and an understanding of basic digital. The understanding of basic digital concepts such as clocking, bit rates, digital thresholds, digital compare, and digital capture are important.

Figure 5.1 shows a block diagram of each of the configurations listed above.

Notice in the figure that the inputs and outputs are specifically separated. This is done purposefully to help illustrate the TX and RX sections. Rather than having separate pins, these signals are often switched internally to the SOC to save on the final pin count. The package size dictates the number of pins that a particular SOC can have, so often internal switches are included in the design to adhere to the pin count requirements.

The internal switching and sharing of various redundant blocks are best illustrated with a picture. Figure 5.2 is a block diagram of a typical Bluetooth radio.

The first thing to notice is that the radio has three main sections: TX, RX, and PLL. Notice that on the antenna side, a switch (RF duplex switch) is used to switch the signal back and forth between the TX section and the RX section of the radio. In reality, this is a good idea for the simple fact that the radio cannot transmit and receive signals simultaneously. Also, notice that the TX and RX mixers share the PLL block. Now, what is Bluetooth and how are Bluetooth devices tested?

5.3 Origins of Bluetooth

Bluetooth began as an open standard project in 1994 by Ericsson in Sweden and was originally named multicommunicator (MC) link—not a very memorable name). The goal was to develop a wireless communication standard that would

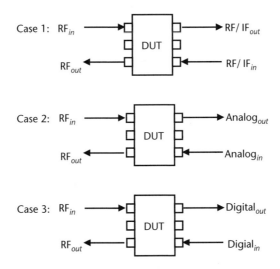

Figure 5.1 Wireless SOC configurations.

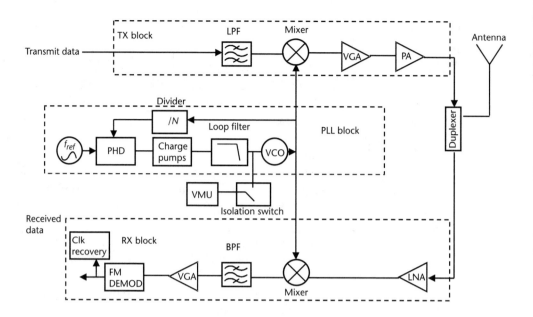

Figure 5.2 Block diagram of a typical Bluetooth radio modem.

support short-range voice and data transfers amongst multiple devices. Four years later, in 1998, four other companies, IBM, Intel, Nokia, and Toshiba, joined with Ericsson to form a special interest group (SIG) to promote the standard and promptly renamed the standard to Bluetooth (a much more memorable name). Today, the promoter group consists of nine companies: 3Com, Lucent Technologies, Microsoft, Motorola, IBM, Intel, Nokia, Toshiba, and, of course, Ericsson. There are also hundreds of associate and adopter member companies. The Bluetooth SIG is driving a low-cost short-range wireless specification for connecting mobile devices.

Bluetooth was coined from the name of a tenth-century Danish king, Harold Bluetooth. The Viking king was credited with uniting Norway and Denmark during his reign. The promise of Bluetooth is seamless interconnectivity among devices (i.e., uniting technologies like the king). Computers, wireless headsets, printers, personal digital assistants (PDAs), mobile phones, and laptops will be able to share files and transfer both voice and data [1].

5.4 Introduction to Bluetooth

Bluetooth is a global wireless standard that in its simplest form is designed to replace cables. The cost must be extremely low, and the devices must be easy to operate. Additionally, the devices must be robust because Bluetooth devices operate in the unlicensed ISM band at 2.4 GHz. The ISM band is reserved for the general use of devices that operate to specifications determined by the various geographical governing bodies such as the Federal Communications Commission (FCC) in the United States and European Telecommunications Standards Institute (ETSI) in Europe.

The ISM band is unlicensed, which means that anyone can operate a wireless device in the band as long as it adheres to the regulations specified in the particular geographical location; other bands require a license (which is expensive) and adherence to government regulations for said spectrum. In contrast to the ISM band, one of the 3G UMTS bands in Europe covers a spectrum from 1,900 to 1,980 MHz. Providers wishing to offer 3G services must obtain an expensive license to do so.

Table 5.2 shows the ISM band allocations versus geography.

In the ISM band you can find a myriad of short-range devices for many applications like wireless local area network (WLAN) applications and, of course, all of our microwave ovens (operating at 2.45 GHz). Bluetooth devices must combat the noisy and overcrowded environment of the ISM band, and it does so by employing three critical techniques to minimize interference from other devices: frequency-hopping spread spectrum (FHSS), short data packets, and adaptive power control [2].

Since I have been working in wireless/RF applications and testing for more than 10 years, I frequently come across some very amusing stories from customers. There was a customer that had installed a Bluetooth network, and it was working just fine for him. But everyday at around noon, the entire network would stop functioning for about 30 minutes. It took several days for him to track down the problem. It turned out that the cafeteria had multiple microwave ovens for employee use. All of the employees were going to the cafeteria at noon every day and using all of microwave ovens simultaneously to warm up their lunches, thus interfering with his network. Bit error rate (BER) tests (discussed later in this chapter) with modulated interferers are designed to catch this exact use case. In general, the Bluetooth network should have been unaffected by the microwave ovens, as the standard is designed to accommodate both operating simultaneously. Perhaps the ovens were outside their leakage specifications or perhaps there were too many ovens in close proximity to the network.

5.5 Frequency Hopping

The history of FHSS dates back to World War II when Hedy Lamarr, the Austrian born actress, and George Antheil, an American composer, copatented an idea to prevent intentional jamming of radar/communication signals. If a signal is being transmitted constantly at a specific frequency, it is a simple matter to interfere with the signal by transmitting at the same frequency. They envisioned that by jumping or "hopping" frequencies faster than the enemy could retune their jamming signal, they could preserve the integrity of the information on the signal. Since the ISM band is already crowded, it is highly likely that multiple devices attempting to

Table 5.2 ISM Band Allocations

Country	Frequency Band (MHz)	Number of Channels
United States	2,400–2,483.5	79
Europe	2,400–2,483.5	79
Spain	2,445–2,475	23
France	2,446.5–2,483.5	23
Japan	2,471–2,497	23

transmit at the same frequency will coexist. To limit interference in the unlicensed ISM band, the FCC regulations place limits on maximum power transmission. The regulation permits a transmit power level only up to 0 dBm. This is really not enough power to ensure reliable operation of the wireless network. To circumvent this challenge and still comply with the regulations, a frequency-hopping spread spectrum (FHSS) technique with a hop speed of 1,600 hops/s is used. Using FHSS, the power level can be as high as 20 dBm, and the range of the wireless network can be extended to 100m. Frequency hopping acts to spread the power across the ISM band and, thus, still adheres to the 0-dBm regulation, provides robust communications, and also acts as a means of security.

You can imagine that a chip manufacturer will desire to test this functionality since it is a critical success factor. There are numerous tests defined in the Bluetooth RF specifications standard that include frequency hopping. To test many of the frequency hopping capabilities, the test system itself must also be capable of frequency hopping. The major hurdle to overcome is that the test system must be able to frequency hop as fast or faster than the SOC device, and this includes any overhead that the test system may have. Additionally, a Bluetooth device will have a trigger pin for the hopping functionality. This pin must be synchronized to the test system.

5.6 Bluetooth Modulation

Bluetooth uses Gaussian frequency shift keying (GFSK). More explicitly, Bluetooth uses 0.5 BT Gaussian-filtered two-frequency shift keying (2FSK), also referred to as binary frequency shift keying (BFSK) at 1 Msymbol/s with a channel spacing of 1 MHz. Since only two frequencies are used, one bit is one symbol (i.e., a 1 indicates a positive frequency deviation (nominally +157.5 kHz) from the carrier, and a 0 indicates a negative frequency deviation (nominally –157.5 kHz) from the carrier). Figure 5.3 shows the amplitude versus time, as well as a constellation diagram plot of 2FSK modulation that is used for Bluetooth. The frequency deviation range is between 140 and 175 kHz.

5.7 Bluetooth Data Rates and Data Packets

The theoretical maximum data rate is 1 Mbps, but due to overhead, the maximum realizable asymmetric data rate is reduced to 723.2 Kbps. This is also a bit misleading because the reverse link has a much lower data rate. Table 5.3 is a summary of the possible data rates for the various packet sizes.

The information is transmitted in a packet in a time slot. Each time slot corresponds to an RF hop frequency. A packet of information can be transmitted in one time slot, three time slots, or five time slots. Naturally, a five-slot packet carries more information than a three-slot packet, which carries more than a one-slot packet. Data high (DH) rate achieves higher data rates by using less error correction in the packets. Data medium (DM) rate achieves a lower bit error rate probability by using more error correction in the packets. A Bluetooth packet is shown in Figure 5.4.

5.7 Bluetooth Data Rates and Data Packets

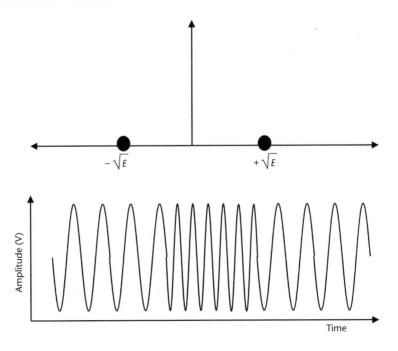

Figure 5.3 2FSK constellation and amplitude versus time.

Table 5.3 Bluetooth Data Rates

Packet Type	Max Symmetric Data Rate (Kbps)	Forward Asymmetric Data Rate (Kbps)	Reverse Asymmetric Data Rate (Kbps)
DM1	108.8	108.8	108.8
DH1	172.8	172.8	172.8
DM3	258.1	387.2	54.4
DH3	390.4	585.6	86.4
DM5	286.7	477.8	36.3
DH5	433.9	723.2	57.6

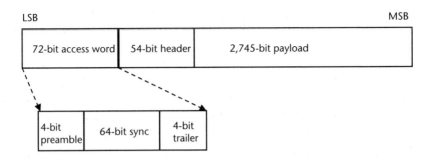

Figure 5.4 Bluetooth packet.

The packet contains a 72-bit access code, a 54-bit header, and a 0 to 2,745-bit payload. The access code is subdivided into a 4-bit preamble, a 64-bit synchronization word, and a 4-bit trailer. This information is important to understand because the various wireless tests that are performed often use or do not use portions of the packet, and the test system must be able to isolate to a specific region in the packet. For example, the initial carrier frequency tolerance (ICFT) is determined by measuring the frequency deviation of the 4-bit preamble of a DH1 packet, so the test system must be able to distinguish among the various subsections of the packet. Another example is the carrier frequency drift, where the drift frequency is determined by measuring the frequency of the payload and using the preamble as the initial reference frequency [3, 4].

5.8 Adaptive Power Control

Recall that regulations limit the transmit power to 0 dBm. If higher transmit power levels are desired, then a spread spectrum technique must be utilized. There are three power classes defined in the Bluetooth specification. Table 5.4 lists the classes along with their corresponding maximum transmit output powers and power-level control range.

Power class 3 is the most common class being adopted by manufacturers for the simple fact that class 3 consumes the least amount of power.

5.9 The Parts of a Bluetooth Radio

We alluded earlier to the three radio sections (TX, RX, and PLL) shown in Figure 5.2 We will now examine each section in more detail, taking a look at the building blocks that are used for each section as it relates to production testing. Notice that the RX section of Figure 5.2 has six individual building blocks (LNA, Mixer, BPF, IF amplifier, FM demodulator, and clock-recovery block). Each individual block has a host of tests that could be required by a manufacturer if the block were an isolated building block (i.e., stand-alone in its own package). The LNA, for example, has numerous noise, power, and S-parameter tests that are described in detail in Chapters 2 and 4. Mixer testing is described in Chapter 2. General filter testing and RF amplifier testing are described in Chapter 4. Modulators and demodulators were introduced in Chapter 2 with a description of some general tests that are critical to modulators and demodulators. The clock-recovery block may or may not be on the wireless SOC. The clock-recovery circuit serves to recover or find the clock so that the baseband chip, which will ultimately be used in conjunction with the wireless SOC in the final application, will know when to sample the valid data. When the clock-recovery block is part of the wireless SOC, it makes testing much simpler. When it is not part of the SOC, then special techniques like oversampling the measured data must be used. Test systems can easily manage this

Table 5.4 TX Power Classes

Power Class	Maximum Output Power (dBm)	Power Control Range (dBm)
1	20	4 to 20
2	4	−30 to 4
3	0	−30 to 0

activity, but the real issue with production testing is test time. Oversampling inevitably increases test time, and the extra test time incurred can be quite substantial. It is possible to implement other solutions to avoid oversampling, but this usually involves specific knowledge of the clock-recovery block and is often not feasible. Clock recovery circuits vary according to the requirements of the standard and the designer's background and experience; thus, they are usually custom in nature. If all of the blocks are integrated into a single unit (as in this example), then the section is called a receiver, or RX for short. A receiver requires a different set of tests in contrast to any one of the individual blocks taken separately. These tests will be discussed later in Section 5.21.

The counterpart to the RX section is the TX section (transmitter). In the TX section of Figure 5.2 there are four blocks (LPF, mixer, VGA, and PA). Again, filter, mixer, and amplifier testing were already mentioned in previous chapters. Also, recall that power amplifier testing is a special case of amplifier testing because it is often a stand-alone block, has thermal issues, is often of a different material, and has high power requirements and efficiency measurements associated with it. The modulator block has not been included in the figure, but it could be off chip, or it could share the modulator from the RX section. The various tests that are performed on a transmitter will be discussed in Section 5.20.

The final block to consider is the PLL. A PLL is essential to any radio, so it is worthwhile to examine the individual blocks to obtain a better understanding of the testing usually performed on a PLL.

5.10 Phase Locked Loop

Bluetooth has explicit specifications that specify the minimum performance parameters for the RF system. However, many of the parameters are theoretical limits and are acceptable only on paper. Many of the specifications do not address the real-world situations that Bluetooth devices can find themselves in [5].

For example, the Bluetooth specification does not specify the synthesizer settling time, but the synthesizer settling time is a key performance factor in any system. This is mainly due to the high overhead of the protocol processor and baseband processing. This places a practical limit of 180 μs on the synthesizer settling time, but most Bluetooth suppliers offer solutions that have synthesizer settling times much better than this [5]. A synthesizer is a PLL, or to be more exact, a PLL is a portion of a synthesizer, and often the two words are used synonymously.

That being said, there are numerous books on synthesizer (PLL) designs, such as integer PLLs or fractional N PLLs, and how they operate. No attempt to describe PLL design methodology will be made since that is out of the scope of this book. If you would like further detailed information on PLLs, take a look at Alain Blanchard's *Phase-Locked Loops* [6]. We will merely discuss the building blocks that are used to build a PLL to tackle the subject from the test engineer's or product/application engineer's point of view.

PLLs come in various flavors, but they essentially have the following blocks: divider(s), a phase detector (PHD), charge pumps, a loop filter (lowpass), and a voltage controlled oscillator (VCO). All of these components are shown in Figure 5.2. Let's discuss each block briefly.

5.11 Divider

A divider divides the input frequency by a programmable integer value to produce a lower "divided" output frequency. As an example, if the input frequency is 100 MHz and the divide value is 2, then the output of the divider will be 50 MHz.

5.12 Phase Detector, Charge Pumps, and LPF

A phase detector has two inputs and one output. A phase detector determines the difference in phase of the two input signals and provides that difference at the output. The charge pumps are usually current sources that either sink or source a current based on the input signal provided to them. A loop filter is a lowpass filter with a specific bandwidth that is specially tuned for a particular PLL application.

5.13 Voltage Controlled Oscillator

A VCO has one input and one output. The input is a voltage signal that can swing according to the operating voltage of the chip, 0V to 3V for example. By changing the voltage, the output tone's oscillating frequency also changes in a linear fashion. For the Bluetooth standard, the VCO needs to cover the entire ISM band. This means that it needs to be tunable across a 75-MHz range. Using the 75-MHz and the 3-V example from above, this means that the VCO has a tuning slope of 25 MHz/V. This parameter is often measured during production. In theory, measuring the tuning slope is straightforward: Apply a voltage and measure the resulting frequency. Repeat this for two different voltages, and now you have two points to make a line and determine the slope. In reality, this measurement is difficult to make due to the high degree of sensitivity of the PLL circuitry. As stated above, the loop filter is specially designed and tuned for a particular application. To measure the voltage, the circuitry must be contacted with a voltage-measuring device, and this perturbs the filter characteristics. So, in reality, extra isolation switches are often required on the load [DUT interface board (DIB)] board if this measurement is important enough to the customer to be included in the final production program [see isolation switch and voltage measurement unit (VMU) blocks in Figure 5.2]. The extra isolation switches allow the measurement to be made while minimizing loop bandwidth perturbations.

5.14 How Does a PLL Work?

In Figure 5.2 the VCO output is fed back to the dividers and is also fed to both mixers in the TX and RX sections. The VCO is used as the LO input to the mixers. The dividers can be programmed by the digital control logic inside the chip. Changing the dividers changes (or divides) the output frequency that is then fed into the phase detector. This is how the carrier frequency is selected. You might envision a microprocessor with the hop frequencies preprogrammed into it. The microprocessor can then quickly and easily hop the frequency by simply reprogramming the dividers. The reference frequency is from a crystal (usually off chip) and is fed into the phase detector. The phase detector detects the phase difference between the divided VCO frequency and the reference crystal frequency. The difference in phase (or phase error) causes the charge pumps either to source or sink current, which after being lowpass filtered is essentially an error signal that is converted to a voltage and used to tune the VCO either up or down in frequency. This entire process operates in a closed loop and iterates upon itself until the VCO has tuned to the desired frequency, thus causing the phase error to go to zero (i.e., the VCO is locked to the desired frequency). This raises the question, How long does it take for the VCO to lock? This tested parameter has several names, such as PLL lock time or synthesizer settling time. We will refer to it as synthesizer settling time, and it is a critical parameter that is measured by wireless SOC manufacturers.

5.15 Synthesizer Settling Time

Synthesizer settling time and synthesizer lock time are synonymous and are often used interchangeably by engineers. In this text, we will use synthesizer settling time. The amount of time required for the synthesizer to settle or lock up is a key performance parameter. However, the synthesizer settling time is not defined in the Bluetooth specification. Moreover, the testing of the synthesizer settling time is usually left completely up to the designer and test engineer/product engineer/application engineer. This means that the engineer must have a very good understanding of the test system resources that are available. The engineer must also have a good grasp of the final big picture regarding the final production test. Keeping it simple and utilizing the resources of the test system is often the best bet for success.

The synthesizer settling time is usually defined as follows: from (a) the time that the synthesizer has received a program instruction to move to a particular channel to (b) the time that the synthesizer has settled to within 10% of the final power of that particular channel. For example, a Bluetooth synthesizer is currently locked to channel 39 and receives an instruction to go to a new channel number (say channel 45). The microprocessor issues the instruction and channel 45 programs the dividers so that a divided frequency sourced from the VCO is phase compared to the reference crystal frequency, and the VCO tunes until the phase error goes to zero.

Since the synthesizer settling time is a key performance factor for any Bluetooth radio, and since suppliers are specifying the settling time in their data sheets, it stands to reason that this parameter needs to be tested in order to ensure a functioning radio in the final application. This is where the jobs of the product engineer, test engineer, and application engineer all become very interesting and critical. In every part of the world, except France, Spain, and Japan, Bluetooth devices have 79 available channels. That means that the synthesizer must be able to settle to any one of the 79 channels within the specified settling time. This implies that all 79 channels of the synthesizer must be tested. Bluetooth devices, by definition, must be very low cost in order to ensure acceptance by the general public [2]. Testing all 79 channels would drive the cost of test per chip too high, so how can all 79 channels be tested to ensure a quality product? The answer is very debatable, and essentially it becomes a trade-off between test coverage (quality) versus device cost (dollars).

One common strategy is to characterize the device heavily in the laboratory until the chip supplier has a high confidence level that the synthesizer settles to all specified channels within the specified settling time and then to adopt the worst-case scenario for production testing. If this strategy is adopted, then physics should dictate that the worst-case settling time would be between the two endpoints, channel 0 to channel 79. Physically, it takes more time for the VCO to slew from channel 0 to channel 79 (or vice versa) than it does for the VCO to slew from channel 0 to any other channel.[1]

5.16 Testing Synthesizer Settling Time in Production

How is such a test implemented in production? Well, the testing strategy must be broken down into smaller segments to determine the test equipment that will be needed. First, the method to measure the frequency must be determined. A simple frequency counter circuit at the output of the VCO/synthesizer could be used, but the overall testing requirements for production testing have to be considered. In most cases, the supplier will want to test the phase noise of the VCO, as well as the synthesizer settling time, since phase noise is another key performance factor. Testing phase noise (see Chapter 8) dictates the use of some sort of digitizing receiver so that a fast Fourier transform (FFT) can be used to obtain the phase noise at a particular offset frequency. The frequency can be measured with a digitizer, but the phase noise cannot be measured with a counter. Since it is desired to minimize the use of switches in a test solution, two common methods for measuring frequency with a digitizer will be discussed: (1) power versus time, and (2) differential phase versus time.

5.17 Power Versus Time

In our Bluetooth example, it was predetermined that the synthesizer settling time from channel 0 (2.402 GHz) to channel 79 (2.480 GHz) must be tested to test the worst-case synthesizer settling time. The test system must utilize a downconverter to intermediate frequency (IF) followed by a tunable digitizer that can provide the narrowband power of the original VCO signal. Then, if the internal local oscillator (LO) of the digitizer is tuned to the same IF frequency, the digitized voltage time samples represent the actual power of the original VCO signal. Figure 5.5 shows an example of the test interface to the DUT.

---

1. In this example, the assumption is that one VCO covers the entire band. In reality, there may be several VCOs and this assumption may not hold.

5.17 Power Versus Time

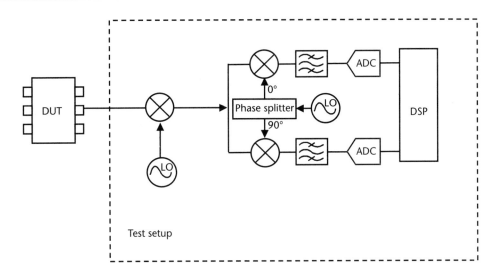

Figure 5.5 Test setup to measure synthesizer settling time.

Why is it important to use a narrow bandwidth? If the bandwidth is too wide, then power in some nearby spurious signal may inadvertently be measured, mistaking that the synthesizer is working when in fact it may not be working at all.

Notice that this method does not directly measure the frequency of the VCO, but rather by tuning the digitizer's LO to the IF frequency, the power that appears at that frequency is measured. In short, this method looks for power at the IF frequency, and if power is found, then it implicitly must be coming from the VCO.

How can this measurement setup be realized? A means to trigger the digitizer to start capturing time samples is required. In addition, the trigger pulse must be sent at the same time that the instructional word to the DUT is sent.

If the digitizer is triggered at the same time the DUT is programmed to change from channel 0 to channel 79, the power (watts) can be calculated from the captured time samples as follows:

$$Power(n) = \sqrt{V_R^2(n) + V_I^2(n)} \quad (5.1)$$

where $V_R(n)$ and $V_I(n)$ represent the real and imaginary portions of a complex voltage waveform, and n is the integer sample number.

A sample frequency should be chosen in order to obtain the narrowband power of the desired signal and to obtain the desired resolution between samples.

$$f_s = \frac{1}{T_s} \quad (5.2)$$

where f_s is the sample frequency and T_s is the resolution between time samples. For example, a sample frequency of 300 kHz implies that the time between samples is 3.33 μs.

$$T_s = 1/300\,e3 = 3.33\,\mu s$$

This means that the worst-case settling time measurement uncertainty is half the distance between sample points.

$$Uncertainty = \frac{1}{2}T_s \qquad (5.3)$$

For the 300-kHz example, the uncertainty is (1/2) 3.333 = 1.665 μs. The engineer must be careful to choose T_s to provide acceptable measurement results. This is especially true if the measurement results are near or within the uncertainty. Let's say the uncertainty is 1.665 μs, the settling time of the synthesizer is specified to be less than 100 μs, and the 90% measurement result point is 101.665 μs. The device may or may not actually be failing the specification. More resolution is required to be certain.

The number of samples to capture must also be chosen to minimize test time:

$$t_{capture} = nT_s \qquad (5.4)$$

For example, using (5.4), 512 samples yields a capture time of

$$512 \times 3.33\,\mu s = 1.70\ mS$$

This is more than enough given that the synthesizer settling time must be less than 180 μs.

Next, a search through the time samples to find the point which deviates from the final average power by 10% must occur. Once the 10% deviation point is determined, call it $n_{90\%}$, multiply that particular point by T_s to obtain the synthesizer settling time.

$$t_{settle} = n_{90\%}T_s \qquad (5.5)$$

where t_{settle} is the synthesizer settling time, $n_{90\%}$ is the index of the 90% power point, and T_s is the sample period. Figure 5.6 is a plot of a typical result using the power-versus-time method.

In the example figure, the 90% point occurs at index $n = 20$. Using (5.5), the synthesizer settling time = 20 × 3.33 = 66.66 μs. Below is power-versus-time pseudocode that determines the 90% synthesizer settling time. The algorithm assumes that the complex voltage samples have been captured at the appropriate IF frequency. The algorithm converts the voltage samples to power, determines the average power of the last k values, and then searches backwards through the power array for the first point that deviates from the average power by 10%. The index of that point is then divided by the sample frequency to obtain the settling time.

```
//****************************************************************

// Example c-code for calculating the synthesizer settling time
using power vs time

//****************************************************************
```

5.17 Power Versus Time

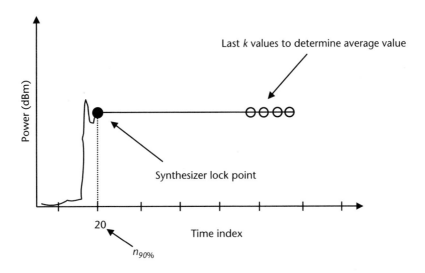

Figure 5.6 Synthesizer settling time result using power-versus-time method.

```
for ( i = 0; i < data_size; i++ ) // Compute the power from the
complex voltage samples

// data_size is the size of the complex voltage time array

{
// Using (5.1)

    voltPower[i] = sqrt( voltReal[i] * voltReal[i] + voltImag[i] *
    voltImag[i] );

    }
// Determine Final Settled Value

    powerLevel = 0.0;           // Initialize powerLevel

    kValues = 5;                // use 5 samples at end of
    array as "settled" power level

    for (i = 0; i<kValues; i++)

        {

    settledPowerLevel = settledPowerLevel + voltPower[numPoints-
    kValues+i];

        }

    // Determine "average" settled power value

    settledPowerLevel = settledPowerlevel / kValues;

    // Determine the 90% Settling Time

    i = numPoints - 1;          // number of power samples

    settleLowPowerLevel = 0.9 * settledPowerlevel; // lower
    threshold value
```

```
settleHiPowerLevel     = 1.1 * settledPowerlevel; // upper
threshold value
```

```
// Search backwards through the array until lower or upper value
is found
```

```
while ((( voltPower[i] > = settleLowPowerLevel) && (voltPower[i]
< =
```

```
settleHiPowerLevel)) && (i > 0)) { i--; }
```

```
// Convert synthesizer settling time to micro seconds
```

```
synthesizerSettleTime = (i + 1) * 1.0e6 / sampleRate;
```

Note: It is important not to search forward in the array; erroneous presettling indexes may be found, resulting in better (faster) settling times, which is undesirable.

5.18 Differential Phase Versus Time

Another method to measure the frequency is based on the same test setup as show in Figure 5.5. Instead of using the time samples to calculate the power in the signal, however, the samples are used to calculate the phase of the signal, and then the derivative is calculated and used to determine directly the frequency of the signal. First, the phase of each individual sample must be calculated.

$$\varphi(n) = \tan^{-1}\left(\frac{V_I(n)}{V_R(n)}\right) \tag{5.6}$$

Again, $V_R(n)$ and $V_I(n)$ represent the complex real and imaginary voltage waveform, and n is the sample number.

$$\varphi(n+1) = \tan^{-1}\left(\frac{V_I(n+1)}{V_R(n+1)}\right) \tag{5.7}$$

$$\varphi(n+512) = \tan^{-1}\left(\frac{V_I(n+512)}{V_R(n+512)}\right) \tag{5.8}$$

Note: Depending on the software package that you are using, take care to unwrap the phase about 2π radians.

Next, the differential phase for any n can be calculated by

$$\frac{d\phi}{d}(n) = \frac{\phi(n) - \phi(n-1)}{T} = 0 \Rightarrow \phi(n) - \phi(n-1) = 0 \tag{5.9}$$

where T is the time step, but since the goal of the analysis is to find when this quantity goes to zero, the T term is canceled. Also, note that there is no 0th differential phase sample, and it is immaterial.

Figure 5.7 shows what a typical synthesizer settling plot might look like for a frequency-versus-time method. Note, the frequency result will be relative to the LO.

5.18 Differential Phase Versus Time

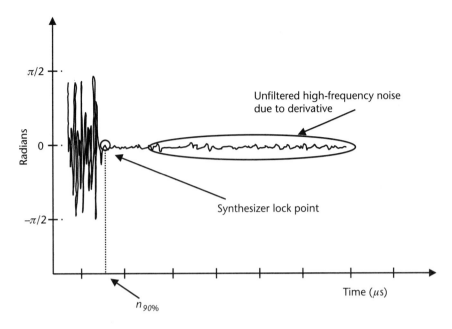

Figure 5.7 Synthesizer settling time using frequency-versus-time method.

Since the LO is tuned to the desired settling frequency, a lock condition will result when the frequency result is equal to zero (i.e., equals the LO).

An important thing that must be considered is that taking the derivative will pronounce any high-frequency noise that may be present in the signal. If this is a problem for the application, then additional digital filtering within the algorithm to suppress the high-frequency noise may be required. Another point to consider is that arc tangents require more processing time than simple arithmetic, so this method may require more DSP time. For the most advanced engineers, look-up tables for arc tangents can be employed to reduce this intensive DSP calculation.

The following is example C code to calculate the differential phase versus time. The code loops through the number of points and determines the phase in radians of each sample (phase[i]). It then determines the phase derivative and unwraps the result about $\pm \pi$. The frequency is then calculated by multiplying the result with the sample rate in hertz and dividing by 2π.

```
//*************************************************************

// Example c-code for calculating the phase derivative

//*************************************************************

for ( i = 0; i < num_points; i++ )        // num_points is the complex voltage array size

{

phase[i] = atan2( Imag[i], Real[i] ); // Determine the phase of each complex sample
```

```
            if (i > 0)

                {

                    phaseDerivative[i] = phase[i] - phase[i-1]; // Compute
                    the Derivative of the Phase

                    if ( phaseDerivative[i] < -M_PI )

                        {

                            phaseDerivative[i] = phaseDerivative[i] + 2*M_PI;
    // Unwrap about 2π

                        }

                    if ( phaseDerivative_[i] > M_PI )

                        {

                            phaseDerivative[i] = phaseDerivative[i] - 2*M_PI;
    // Unwrap about 2π

                        }

            // Compute the frequency in kHz
    frequency[i] = (sampleRate) * ( phaseDerivative[i] ) /
    (1000*(2*M_PI));

                }

    }
```

This method requires that the RF receiver (IF downconverter with tunable digitizer) is time synchronized or phase locked with the digital system. The digital system and RF receiver system both require a common time base. This is usually implemented with a reference clock or master clock. If their clocks are not synchronized (i.e., phase locked), the resulting measurement will be erroneous.

5.19 Digital Control of an SOC

Most SOC devices utilize a three-wire serial protocol interface (SPI) to program different modes, power levels, channels, and so forth, of the chip. A typical three-wire SPI timing diagram is shown in Figure 5.8. There are three control lines, namely clock, data, and enable. The enable line is used to enable (validate) the clock and data lines. Either a constant high or a constant low will be used to enable the data and clock. The SPI data is the actual word that programs the desired function to the SOC and is typically a 16-bit to 32-bit serial word. The frequency of the SPI data is typically in the 1- to 10-MHz range. The SPI clock is completely independent of any other clocks that may be required by the application; it always has a 50% duty cycle and runs at twice the frequency of the SPI data. The data is clocked into the application on either a rising or falling edge of the SPI clock; thus, the SPI clock must run at twice the speed of the data.

5.20 Transmitter Tests

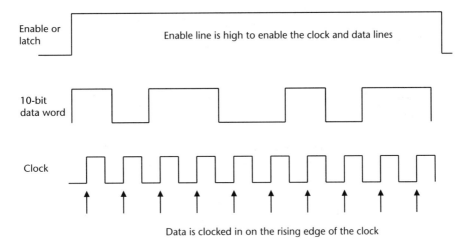

Figure 5.8 Typical three-wire SPI timing diagram.

One important programming aspect already mentioned is the ability to control the frequency. It is important to be able to program the SOC to hop to different frequencies. Other important programming aspects include power and gain control, switching among the various modes (i.e., transmit, receive, power down, or stand by), switching to different bands (i.e., 802.11a, 802.11b, or 802.11g, in the case of a WLAN SOC), and test modes. The manufacturer's chip specifications describe the functionality of the digital control in detail, and each SOC will be slightly different. But the digital control requires digital testing, and this can be generalized by the following:

- The upper and lower voltage thresholds are tested.
- The timing, including rise times and fall times, is tested.
- The maximum frequency at which valid data can still be received is tested.

These are very basic tests by digital standards, but the point is that the test system must have digital capabilities that must be synchronized with the mixed-signal and RF functions of the test system.

5.20 Transmitter Tests

Recall the various TX blocks in Figure 5.2, namely LPF, mixer, IF amplifier, PA, and modulator. We mentioned that many of the associated measurements (for example, filter roll-off, power compression, S-parameters, and phase imbalance) are covered in detail in other chapters. Loosely speaking, we can consider all of those measurements as parametric measurements. Now that all of the blocks are integrated into a single SOC, it is desirable to perform many of those parametric measurements across the SOC (for example, transmit CW output power with gain control). But it is also desirable (usually mandatory) to perform functional tests on the SOC to ensure that it will function in the end application. The remainder of this section will concentrate on the functional TX tests that are often performed on Bluetooth SOC devices. There are four challenging key functional TX tests that manufacturers may require for production. Table 5.5 lists these tests with common test parameters.

The list in Table 5.5 is taken directly from the Bluetooth standard [3, 4]. Often, variants of these parameters are required by the SOC manufacturer. There are three different payload types called out by the standard: PRBS 9, 11110000, and 10101010. Each type stresses the modulator differently and is specifically chosen for a particular measurement. PRBS 9 stands for pseudorandom bit sequence, has a periodicity of $2^9 - 1$, and is intended to approximate live-traffic data. During normal operation, the data is completely random, so a PRBS 9 pattern simulates normal operation to produce the same spectral distribution. The 10101010 pattern is used to test the modulator's filter. The 11110000 pattern is used to test the Gaussian filtering.

5.20.1 Transmit Output Spectrum

Bluetooth operates in the ISM band, so it must comply with both inband spurious emissions (within the ISM band) and out-of-band spurious emissions (outside of the ISM band) to ensure noninterference with other ISM and non-ISM devices. The transmit output spectrum of the device is compared to the inband mask shown in Table 5.6 and the out-of-band mask shown in Table 5.7.

Where M is the channel number (channel 45 for example) and N is the adjacent channel that is being measured.

The device is programmed to a specific channel in the TX mode. A digital PRBS 9 pattern is fed into the DUT. The modulator will respond to the pseudorandom pattern to produce a modulated RF output. The test system must capture this

Table 5.5 TX Test Parameters

Transmitter Test	Frequency Hopping	Payload Data
Transmit output spectrum	Off	PRBS 9
Modulation characteristics	Off	11110000,10101010
Initial carrier frequency tolerance	On or off	PRBS 9
Carrier frequency drift	On or off	10101010

Table 5.6 Inband Spurious Emissions Mask

Frequency Offset	Transmit Power
$M \pm 500$ kHz	−20 dBc
$|M - N| = 2$	−20 dBm
$|M - N| > 3$	−40 dBm

Table 5.7 Out-of-Band Spurious Emissions Mask

Frequency Band	Operating (dBm)	Idle (dBm)
30 MHz to 1 GHz	−36	−57
1 to 12.75 GHz	−30	−47
1.8 to 1.9 GHz	−47	−47
5.15 to 5.3 GHz	−47	−47

modulated RF signal and compare it to the mask in either Table 5.6 or 5.7 for compliance. Figure 5.9 is a typical inband result.

The figure highlights that the SOC device barely passes the first listed specification, where with $M \pm 500$ kHz, the signal must be attenuated by 20 dB. This is also called the 20-dB bandwidth (BW) test. The manufacturer may concentrate on a particular area of a measurement if he is confident that he can guarantee that the rest of the specifications for that measurement are well within their limits. The 20-dB bandwidth measurement requires considerable test time as it is defined by the Bluetooth specification, so, naturally, chip manufacturers must develop testing strategies that ensure compliance, while minimizing test times.

Examine the ideal 20-dB bandwidth graph shown in Figure 5.10. In the ideal case, the 20-dB bandwidth can easily be determined by first searching for the peak power; let's call it $Power_{peak}$. From the $Power_{peak}$ point we would then search left and right for points with 20 dB less power; let's call them $f_{20\,dBc\_Lower}$ and $f_{20\,dBc\_Higher}$, respectively. The 20-dB bandwidth is then

$$BW_{20\,dB} = f_{20\,dBc\_Higher} - f_{20\,dBc\_Lower} \quad (5.10)$$

where $BW_{20\,dB}$ must be less than 1 MHz.

Unfortunately, to acquire a nearly ideal curve would require seconds of test time. SOC costs demand that the complete test program (including all tests) for a Bluetooth SOC require just a few seconds.

If a spectrum analyzer is used with the same testing methodology and with a minimum test time, then a curve similar to that shown in Figure 5.11 is obtained.

This is obviously unacceptable as there is no reliable way to determine the peak power or the 20-dB crossings with any consistency. However, the result can be improved by tuning the measurement based on the specific requirements. Let's assume the approach of capturing the complex voltage time samples where the

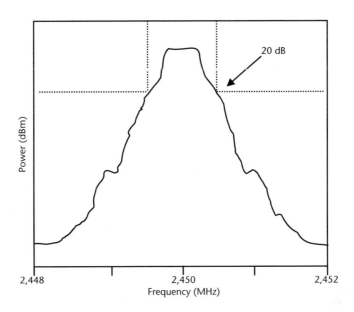

Figure 5.9 Typical Bluetooth transmit spectrum.

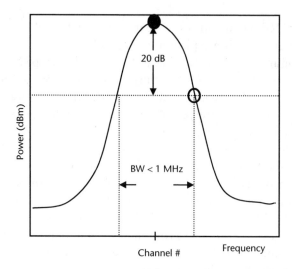

Figure 5.10 Ideal Bluetooth 20-dB bandwidth plot: average power versus frequency.

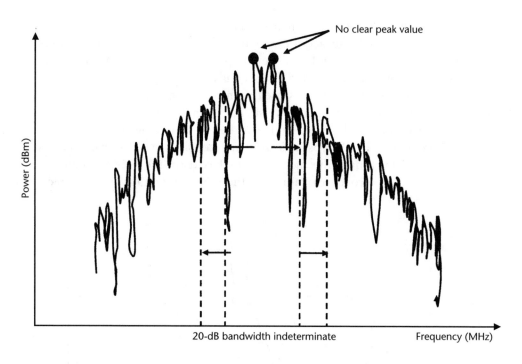

Figure 5.11 Nonideal Bluetooth 20-dB bandwidth plot: average power versus frequency.

output signal has been downconverted to IF. The complex FFT of the samples will provide the output spectrum. A number of FFTs must be taken, and the maximum point for each index point across all of the FFT traces must be recorded to create a worst-case maximum (in spectrum analyzer terms, this is called a *max hold*). A bandwidth of at least 1 MHz must be captured, but the number of time samples and the number of FFTs are arbitrary, meaning, we need to use enough to provide a reliable and repeatable result. If too few samples are used, the result is not reliable, but the test time is acceptable. If too many samples are used, the result is reliable, but the test time is unacceptable. The absolute certainty of the measurement is constrained by the FFT resolution. Let's assume that a 2-MHz capture bandwidth is used and the desired frequency resolution is 1 kHz. The number of points for the FFT is then

$$N = \frac{BW}{Freq\ Resolution_{desired}} = \frac{2e6}{1e3} = 2,000 \tag{5.11}$$

where N is the number of FFT points, and BW is the bandwidth.

Because FFTs execute much faster when powers of two are used, 2,048 points are used. Finally, the number of FFT blocks needed to obtain a repeatable result must be determined. This is an experimental venture where the test engineer must evaluate the measured results for different numbers of FFT blocks versus standard deviations until he has obtained a repeatable result with a low standard deviation. Figure 5.12 is a pictorial representation showing 8,192 points broken into four FFT blocks of 2,048 points each and how the standard deviation will improve with the additional points and additional FFT blocks (i.e., 4,096 points yields σ_2, which is better than 2,048 points which yield σ_1).

Figure 5.13 shows the max hold results of the power-versus-frequency spectrum using four FFT blocks of 2,048 points each of the time samples of the IF's complex voltage waveform. Notice that the result is much improved versus Figure 5.11.

As a comparison, the test time for this implementation will be in the tens of milliseconds versus the original test time of a few seconds.

5.20.2 Modulation Characteristics

Modulation measurements check the modulator as well as the stability of the LO. Since both can be affected by noise problems, these tests were developed to check functionality. If the modulator is not functioning properly, then the frequency

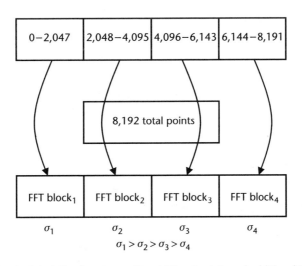

Standard deviation improves with additional points and additional FFTs

Figure 5.12 Standard deviation improvement with additional FFT blocks.

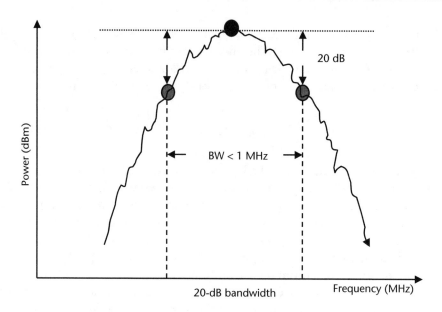

Figure 5.13 Improved Bluetooth 20-dB bandwidth results using four FFT blocks (2,048 points each).

deviation could be incorrect, loosing all of the data. There can also be a pulling effect on the VCO from the power supply, which pulls the carrier frequency from its nominal position. This can also cause the complete loss of data. Finally, the VCO may drift over time so that the last data bits in the packet can be dropped.

The modulation characteristics test is a frequency deviation measurement that tests both the modulator performance and the premodulation filtering. The test system must have vector signal analyzer capabilities because both phase and symbol information is needed for the computations.

Two vector sequences (vector meaning a digital bit sequence), 11110000 and 10101010, are used. The first vector is sequenced over 10 packets, and the frequencies of each of the 8 bits are measured and averaged together. Then the maximum deviation for each bit is computed, as is the average of the maximum deviation. The entire procedure is then repeated for the second vector. The deviations must be between 140 and 175 kHz. The test system must be able to frequency demodulate the captured signal, or the test system must be able to provide the complex time domain of the captured signal. The signal processing can then be performed on the host computer or by the DSP of the digitizer if one is available. As is evident from the description, this test is a DSP-intensive activity and care should be taken to minimize the test time.

5.20.3 Initial Carrier Frequency Tolerance

The accuracy of the TX carrier frequency must be checked. Imagine that the Bluetooth SOC is hopping from one frequency to the next. It takes a certain amount of time for the VCO to jump (or slew) to each frequency, and this is measured by the settling time, which is discussed above. Now, assuming that the synthesizer has just settled, it immediately begins transmitting data. But there is finite accuracy between the programmed channel frequency (desired) and the frequency of where the VCO actually is (the initial frequency). This finite difference is what is measured by the initial carrier frequency tolerance (ICFT). The ICFT result must be less than 75 kHz, and it measured on the first 4 bits (the preamble).

The ICFT test uses a standard DH1 packet with PRBS 9 data. The preamble should be 1010. The captured signal must be frequency demodulated, and the frequency offset of each preamble bit is averaged together. This measurement can also be performed with frequency hopping off.

5.20.4 Carrier Frequency Drift

Bluetooth specifications require a symbol timing accuracy of ± 20 ppm (parts per million). The baseband crystal must be accurate across all operating conditions, temperatures, and the life of the product. With this information a worst-case phase error can be calculated from the longest packet (DH5). A DH5 packet is five slots long, which is 2,870 μs. Thus, the worst-case phase error is

$$40 / 1{,}000{,}000 \times 2{,}870 \ \mu S = 0.12 \ \mu S$$

This is less than one eighth of a symbol period. This leads to the topic of frequency drift. As a Bluetooth SOC is transmitting a packet or series of packets, the synthesizer will slowly drift in frequency. If the synthesizer drifts too far during transmission, then the successive data bits will be lost or dropped during demodulation by the receiver. The maximum drift for one time slot is ± 25 kHz, for three time slots is ± 40 kHz, and for five time slots is ± 40 kHz. The carrier frequency drift measurement is designed to ensure compliance. A 10101010 vector should be used in the payload. The frequency of each of the preamble bits (see Figure 5.4) is measured and integrated; this provides the initial carrier frequency (see Section 5.20.3). Next, the frequencies of each successive 4 bits in the payload are determined and integrated. The frequency drift is then the difference between the average frequency of the preamble bits and the average frequency of any 4 bits of the payload. The maximum drift rate can also be measured and is defined as the worst-case drift between any two successive 4-bit sequences. The maximum drift rate must be less than 400 Hz/s. Figure 5.14 shows a carrier drift measurement where the carrier has drifted 25 kHz.

Again, a test system with vector signal analyzing capabilities is needed to measure the phase and symbol information properly. Creative methods are sometimes used by manufacturers to provide the equivalent results using less DSP horsepower to obtain faster test times. For example, if the Bluetooth SOC has a certain carrier drift specification, then one could perform a "go, no-go" frequency drift test. To do that, the signal of interest is captured to obtain the downconverted complex voltage IF waveform with enough samples to span the time interval of interest (4 ms, for example). An FFT on the beginning samples to determine the initial frequency is performed, and a second FFT on the remaining samples is performed. The maximum drift is then the difference between the initial frequency and the largest frequency of the second FFT. Where exactly in the sequence the worst-case frequency drift occurred cannot be determined with this method, but this information may not be important enough to the manufacturer to pay for it with the additional test time

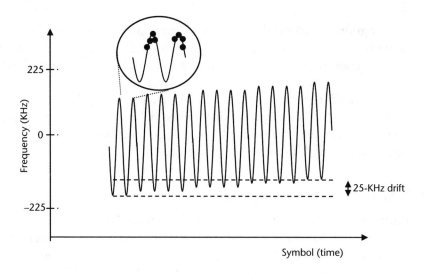

Figure 5.14 Carrier drift of a Bluetooth signal.

that would be required. Using this method, production managers can determine if the device passes or fails its specification and that is usually enough.

5.20.5 VCO Drift

The above carrier frequency drift tests both the modulator's ability to modulate and the drifting of the VCO simultaneously. However, it is also possible to test the drift of the VCO directly without checking the modulator. Testing the VCO drift frequency directly is less DSP intensive because software demodulation to recover the data bits is not required. Thus, this method has test-time advantages, and testing the drift frequency directly may satisfy the production specifications. The test setup required is equivalent to the synthesizer settling time's test setup. Utilizing a differential phase approach (described previously), in many cases it is possible to test the synthesizer settling time and frequency drift simultaneously, thus further reducing test time. Figure 5.15 is a plot of both synthesizer settling time and frequency drift, where a frequency-versus-time method has been used.

5.20.6 Frequency Pulling and Pushing

Frequency pulling and pushing are two common synthesizer tests that are similar to frequency drift. Many wireless SOC devices can operate in a phase-locked or phase-unlocked mode. Switching between the modes will cause the synthesizer to pull the frequency slightly. This parameter can easily be tested utilizing the same methods that are used to obtain the synthesizer settling time and frequency drift.

Variations in the power supply voltage to the wireless SOC device have the effect of pushing the frequency because the PLL and voltage regulator are fighting each other. The test setup required to test this parameter is slightly different in that the ability to change the device power supply voltage dynamically during the test is required, and it must be time synchronized to the test system. With the proper test setup, frequency pushing, frequency pulling, lock time, and frequency drift can be

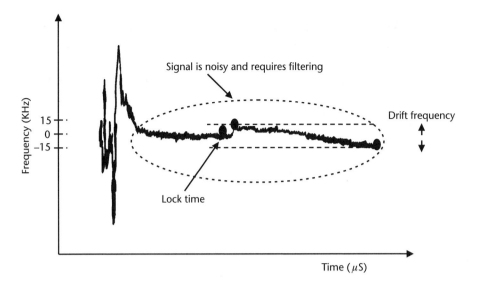

Figure 5.15 Synthesizer settling time and drift using frequency-versus-time method.

tested simultaneously (with one single digitizer capture), thus further reducing production test time.

For example, consider a digital word that is sent to the DUT to program the device to a specific channel. Simultaneously, the test system's RF receiver can be triggered to begin acquiring time samples. Other digital words can subsequently be sent from the test system to the DUT to program drift, pushing, and pulling conditions to the wireless SOC device. The test system has captured one array that can be parsed into the individual lock time, drift, pushing, and pulling segments, which can then be processed or filtered to obtain the desired result. Figure 5.16 is a graph showing the synthesizer settling time and frequency pulling that occurs when the PLL is suddenly unlocked.

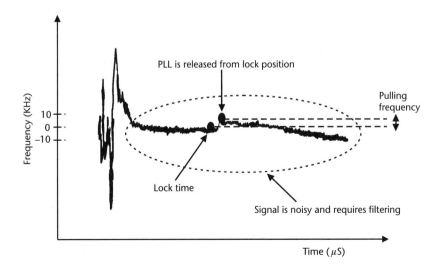

Figure 5.16 Synthesizer settling time and pulling, using frequency-versus-time method.

Below is pseudocode for determining the drift frequency, pushing frequency, and pulling frequency. The settling time, pulling time, drift time, and pushing time indexes have been predetermined for a "go, no-go" test. Each index is windowed to reduce the noise that is inherently enhanced by determining the phase derivative. After that, the resultant values are subtracted from each other to determine the specified relative values.[2]

```
//*************************************************************

// Rolling Window Average is used to smooth the data
// of the Synthesizer lock time, Pulling frequency, and Drift
frequency arrays

//*************************************************************


for ( i = locktime-2 * halfWindowSize; i < = locktime; i++ ) // Create a window size

    {

    // Determine average value inside window

      movingWindowAverageValue + = frequency[i];   }

    // Initialize min and max values

    minLockValue = maxLockValue = movingWindowAverageValue;

    // Slide across window

    for ( i = locktime-halfWwindowSize+1; i < = lock-time+halfWindowSize; i++)

        {

            movingWindowAverageValue - = frequency[i-halfWindowSize-1] +

    frequency[i+halfWindowSize]; // Re-compute average as you slide

        if (movingWindowAverageValue > = maxLockValue) maxLockValue =

    movingWindowAverageValue; // Update maximum

        if (movingWindowAverageValue < minLockValue) minLockValue = movingWindowAverageValue; // Update minimum

        }

    maxLockValue = maxLockValue/2.0/halfWindowSize;
```

2. Windowing code portion written by Ashish Desai.

5.20 Transmitter Tests

```cpp
        minLockValue = minLockValue/2.0/halfWindowSize;

        // Slide across window

        for ( i = pullindexlow - 2 * halfWindowSize; i < =
        pullindexlow; i++)
{
                averagePullingLow + = frequency[i]; // Determine
                average pulling low frequency

                }

        // Slide across window

        for ( i = pullindexhigh; i < = pullindexhigh + 2 *
        halfWindowSize; i++)
{
                averagePullHigh + = frequency[i]; // Determine
                average pulling high frequency

                }

        // Compute the Pulling Frequency

        pullingFrequency = (averagePullHigh - averagePull-
        Low)/2.0/halfWindowSize;

        for ( i = drifttime; i < = drifttime + 2*halfWindowSize;
        i++)
            {
              averageDriftFrequency + = frequency[i];

              }

        // Compute the Drift Frequency

        driftValue = (averageDriftFrequency - averagePull-
        High)/2.0/halfWindowSize;

        cout << "Min Lock = " << minLockValue    << endl;

        cout << "Max Lock = " << maxLockValue    << endl;

        cout << "Pulling = "    << pullingFrequency << endl;

        cout << "Drift    = "       << driftFrequency    << endl;


//****************************************************************

        // Rolling Window Average used to smooth the data

        // for the Pushing Frequency array (see phase derivative
        versus time pseudo code)
```

```
//*****************************************************************
        highIndex = lowGuessIndex * sampleRate);
        lowIndex = highIndex + HighGuessIndex * sampleRate;
        for ( i = highIndex; i < = highIndex + windowSize; i++ )
            {
// Compute the High Side Pushing Frequency
                pushingHighFrequency + = frequency[i];
            }
        pushinghighFrequency = pushinghighFrequency / windowSize;
        for ( i = lowIndex; i < = lowIndex + windowSize; i++ )
            {
            // Compute the Low Side Pushing Frequency
                pushingLowFrequency + = frequency[i];
}
        pushingLowFrequency = pushingLowFrequency / windowSize;
        pushingFrequency = (pushingHighFrequency - pushingLowFrequency);
        cout << "Pushing Frequency: " << pushingFrequency << endl;
```

5.21 Receiver Tests

What is required to test the functionality of a receiver? Conceptually, the testing concept is simple and can be compared to a person's ear. An analogous question could be, How is a person's ability to hear tested? You can imagine that you are in a room full of people and trying to have a conversation with someone in the room. Your ability to hear the other person depends on three things:

1. How clearly does the other person speak (i.e., how well does he transmit)?
2. How sensitive are your ears?
3. How loud are the other people in the room?

The TX tests that were previously discussed take care of item 1. Item 2 can be tested with a simple sensitivity test that would be analogous with having your ears tested by a doctor. A simple sensitivity test (ear examination) has some value, but it does not reflect the real environments that we find ourselves in. As an example, contrast the differences between trying to have a conversation with someone at a movie theater full of people (before the movie has started) versus trying to have a conversation with someone at a nightclub providing live entertainment. Both locations have approximately the same number of people. However, it is quite easy to have a conversation with someone in the theater, while it may be much more difficult to have a conversation with someone at the nightclub. The more people that are talking around you, the more difficult it is to hear the intended person. The constant increase of the noise is synonymous with the term *noise floor*. A theater, before the movie has started, is generally a quiet place (i.e., has a low noise floor). A nightclub with live entertainment is a much louder local (i.e., has a higher noise floor). At the nightclub the other people are interfering with the intended conversation. At some point, as the noise level increases, words from the conversation will begin to be misunderstood or dropped, but the conversation may still be able to be followed. However, eventually, as the people or music get louder and louder (the noise level or noise floor increases), too many words will be missed and the conversation can no longer be understood. The missed words are analogous to misinterpreted digital information, or bit errors, and just as we have the ability to understand a conversation even though a few words are lost, so does a wireless device have the ability to properly decode a transmission even though it encounters a few bit errors.

The RX tests are designed to test the simple case of item 2, as well as the harsh environment case of item 3. In the real environment, the wireless device will receive a signal from a nearby transmitter, and the receiver must be able to demodulate the received signal correctly in the presence of other unwanted signals from other transmitters. In the case of a Bluetooth receiver, it must be able to reject both inband as well as other out-of-band signals that might be in the area.

In the case of a multiband WLAN device, multiple modulation formats and multiple interfering signals across different bands may be needed to test the receiver. More generally, a receiver must be able to correctly identify and demodulate a desired signal at some minimum received power level in the presence of other CW or modulated signals at predefined interfering power levels. For a Bluetooth receiver of an SOC device, a bit error rate (BER) measurement is used to determine the quality of the receiver.

5.21.1 Bit Error Rate

Let's take a closer look at the receiver portion of the Bluetooth block diagram (Figure 5.17).

During the receive process, the device receives a GFSK modulated RF signal in the industry, science, and medicine (ISM) band. The signal is first amplified by the LNA, before being downconverted by the mixer. After being downconverted, the signal is then filtered and amplified again before being fed to the frequency demodulator. The demodulator inside the receiver must sense this signal and demodulate it correctly to produce the original digital bits. It is possible to test each individual block of the receiver chain, but bit error rate (BER) is more effective and is one of the most common tests used to verify that the receiver is functioning properly. BER is defined as

$$BER\% = \frac{Total\ number\ of\ bad\ bits}{Total\ number\ of\ transmitted\ bits} \times 100 \qquad (5.12)$$

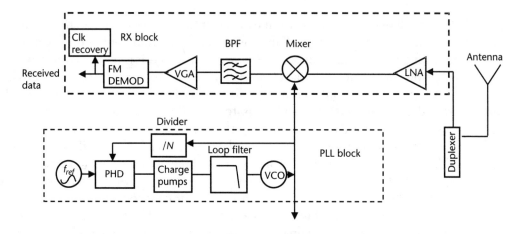

Figure 5.17 Receiver block diagram of a Bluetooth modem.

The concept is quite simple, but rather challenging to implement in a production environment. The ability of a digital vector (data vector) or I/Q generator to modulate an RF signal generator is needed. The ability to control the desired signal's power level and the interfering signals' power levels very accurately is also required. The wireless SOC chip will receive the modulated signal and demodulate it to produce the original digital sequence as an output. In the final application, this output will go to the baseband processor. However, for BER testing, the test system must compare the output digital bit sequence against the original digital bit sequence. See Figure 5.18 for a block diagram of the test setup.

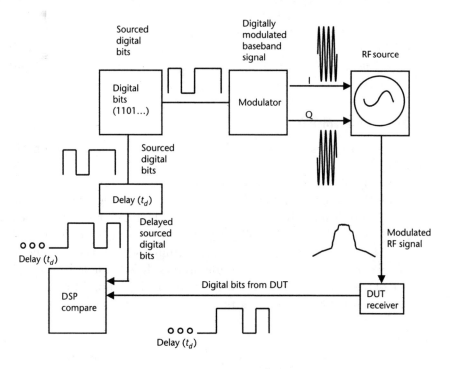

Figure 5.18 Basic BER test setup block diagram.

5.21 Receiver Tests

The first big problem that should jump out at you is, How is the synchronization of the input and output digital sequences performed? There is an obvious delay of the digital information progressing through the system's RF source to the device, and finally through the device, before appearing at the output; it is represented by t_d (time delay) on the figure. For most systems t_d is a few microseconds, but already that poses a challenge. The spacing between bits for Bluetooth is 1 μs, so the delay needs to be calibrated, before comparing the two bit sequences.

5.21.2 Bit Error Rate Methods

First, a testing method that produces a repeatable time delay through the test system must be found. If the delay time through the RF source is repeatable, then the challenge is reduced to one of simply measuring the time delay of when the test is started to when digital data is noticed at the output pin.

Determining the time delay is the main issue that must be resolved. In the next few sections, various implementation methods, all of which address the time delay issue, are discussed. There is a plethora of implementations for BER, and all of them center on the idea of determining the time delay.

5.21.3 Programmable Delay Line Method (XOR Method)

The fastest BER method would be to implement using hardware specifically designed for a particular application. As an example, Figure 5.19 shows a test setup using hardware consisting of a programmable delay line and an exclusive or (XOR) comparator. The XOR truth table is given in Table 5.8. From Table 5.8, one notices that a bad bit will be counted if and only if the two inputs to the XOR comparator are different.

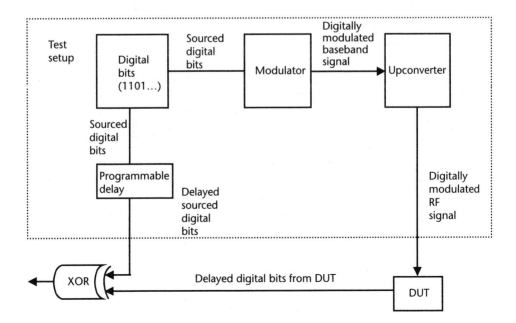

Figure 5.19 BER test setup using programmable delay line with XOR comparator.

Table 5.8 XOR Truth Table

Input 1	Input 2	Output
0	0	0
0	1	1
1	1	0
1	0	1

The idea is to determine the delay time and then to program that value into the programmable delay line. At that point, the XOR comparator is then synchronized and would correctly compare the transmitted digital pattern with the received demodulated digital pattern. The comparison is occurring in real time, so the test time is limited only by the bit speed and number of bits used for the BER test. For example, if the bit speed is 1 Mbps and 50,000 bits are used for the BER test, then the theoretical minimum test time is 50 ms. If there are no errors, then the comparator will always register a TRUE. The output of the comparator can go to a register, for example, and after the test is completed, the register can be queried and the BER result determined. This method works fine as long as there are no errors in the synchronization word, or if errors in the synchronization word are being used to calculate the BER. The Bluetooth specification, however, allows errors in the synchronization word (see Figure 5.4) [3, 4]. Thus, this particular hardware solution, while being very fast, would not discount bit errors in the synchronization word. Bit errors in the synchronization word would show up in the BER result and be inaccurate in its accordance with the Bluetooth specification.

5.21.4 Field Programmable Gate Array Method

Another common method is that of using a field programmable gate array (FPGA) at the output of the device under test. With an FPGA the synchronization error problem can be eliminated. The delay time is programmed into the FPGA and since the location and length of the synchronization word is predefined by the standard, the FPGA is also programmed to ignore particular bit errors in the synchronization word. Figure 5.20 is a block diagram of what an FPGA solution would look like.

This method does have the disadvantage of being more difficult to debug due to the source code being contained inside the FPGA itself. In addition, this method does not lend itself easily to the concept of multisite testing. For each site, another FPGA would be required, and the site's path delay time would have to be characterized for each individual FPGA. For test volumes that require a few test systems, the FPGA solution is probably an acceptable solution, but for test volumes requiring tens of test systems, the solution could prove not to be production worthy. Additionally, the FPGA source code does not have a high reuse factor, meaning the solution cannot be leveraged easily for other products. Thus, the development time (time to market) of the next wireless SOC production solution will not be shortened.

5.21.5 BER Testing with a Digital Pin

A third possible method to measure BER is to capture the DUT's digital output with a digital pin of the test system. Figure 5.21 shows a BER setup utilizing a test system's digital pin.

5.21 Receiver Tests

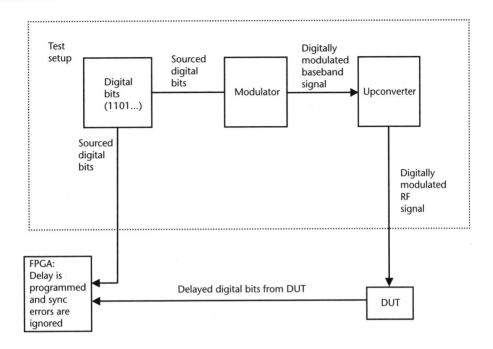

Figure 5.20 BER test setup using FPGA.

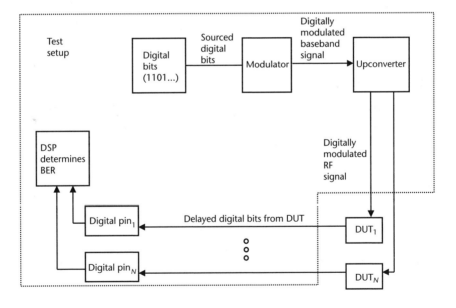

Figure 5.21 BER test setup using digital pin.

This has the advantage of being easily scalable for parallel measurements to as many devices as there are test system digital pins. Another advantage is that the time delay need not be known. By utilizing the digital capture memory of the digital pin, all of the bits are captured and the information is transferred back to the workstation, or PC. The workstation processes the data and determines the BER result for each parallel site. This method is easy to debug and nearly all ATE vendors offer digital pins with some sort of digital capture memory as a standard feature of their product.

The disadvantage of this method, which makes its test-cost prohibitive as a production-test solution, is that it is extremely slow in comparison to the other methods. All of the captured digital data must be transferred back to a workstation or PC, which requires considerable test time (especially for large bit-length sequences, i.e., 100,000 bits). In addition, a software algorithm must be employed on the workstation to process the data (much the same as the FPGA method) to determine the BER.

However, the versatility of this BER method, as well as the low cost of digital pins and the attraction of scalability, has enticed the ATE vendors to improve the test time of this method. There are ways to improve upon this method, so that the low test time of the delay line method can be approached. Digital pins have the ability to compare digital data versus a predetermined vector that has been prestored into the memory of the digital pin. This is similar to the delay line method, but without the ability to determine the delay time. If the delay time can be reliably determined during production, then the digital pin method will have nearly the same production test time as the delay line method.

To determine the delay time, a short training sequence can be used. The training sequence can be captured, transferred to the workstation, and processed to determine the delay time. Another option is that the length of the training sequence can dynamically be changed during runtime, while the BER is continually running. The length of the training sequence corresponding to the lowest BER result will correlate to the time delay required for that particular DUT.

5.21.6 BER Measurement with a Digitizer

Yet another method is to use a digitizer to perform the BER measurement. Many of today's digitizers have both digital inputs and analog inputs. The digital inputs can be utilized to perform BER testing. Most digitizers have sampling rates that are much greater than the 1-Mbps requirement of Bluetooth devices, which make them good candidates to perform these BER tests.

Additionally, many wireless SOC devices do not have the clock-recovery circuitry integrated into the wireless device. This poses no problem for the end application, because the clock-recovery circuitry is on one of the other chips, but this causes a problem for BER testing. Without the clock-recovery circuitry, the test system cannot be triggered by the SOC device. Without the ability to trigger the test system from the SOC device, alternate BER testing methods must be implemented, which are more time-consuming. The test system must then provide its own clock to trigger itself to capture or sample the data. This creates yet another time-delay issue that must be resolved. One possible solution is to oversample the bits and perform a correlation to determine the time delay. This can be time-consuming and care must be taken to minimize the test-time impact.

A digitizer that is capable of sampling at several times the actual bit rate may be required, because of the oversampling rate that will be needed to determine the optimum position within 1 bit. When oversampling is required, improvements in the BER accuracy can be had by masking portions of the oversampled bits. For example, if the oversampling factor is 8 times, masking 1 to 3 bits on the outer edge of each sampled bit may improve the BER by reducing perturbations caused by jitter. In addition, a faster digitizer affords the ability to catch up during the acquisition phase so that test time can be minimized.

For example, if the digitizer runs at 40 Mbps, the digitizer can acquire the Bluetooth output data rate of 1 Mbps until it fills up its buffer memory with enough data to warrant a search for the synchronization word. Once the digitizer has enough bits in memory to search through, it can begin searching for a match on the synchronization word while capturing the rest of the data. In this example, the digitizer can search for 39 cycles compared to the 1 cycle that it needs to capture the next data bit.

This method complicates the test solution, but it has some serious advantages. First, since it is based upon a software solution, the sampling-rate, synchronization lengths, number of bits to capture, packet lengths, and so forth can easily be changed without modifying the hardware. Thus, the reuse factor is high, and this method can be modified for other data-rate schemes, so long as it is supported by the sampling rate of the digitizer. Second, the digitizers normally have multiple digital input pins, so parallel testing can be implemented to reduce the cost of test. With the low cost pressures on Bluetooth and other wireless SOC devices, it is especially important to recognize and use the parallel-digital nature to implement a parallel-testing solution. Lastly, using a digitizer contains the complete solution within the test system, and no extra components are required on the load board or test board. This is increases the test solution's reuse factor.

Table 5.9 compares the four BER methods by throughput, complexity, reusability, hardware cost, maintainability, and ease of development.

The throughput of the XOR and digital pin solutions will be very high because both of them are strictly hardware solutions and require little to no postprocessing of the captured data. The FPGA and digitizer solution both must postprocess the data, which will require additional test time. The XOR and FPGA solutions are more complex because they both require that extra hardware components be placed on the load board or DUT interface board (DIB). Therefore, although they have high throughput, they have low reusability because, for the next device, the engineer will have to go through the process again. In contrast, the digital pin and digitizer solutions have higher reusability because they are built-in to the test system, and the software is fully integrated as well. The hardware cost of the FPGA solution is higher, again due to the extra hardware components. Lastly, because the digital and digitizer solutions are fully integrated into the test system, they score higher for maintainability. Imagine multiple testers in production running a particular solution. If a digital pin fails, it is easy and cheap to replace. However, if an FPGA or XOR fails, it is more difficult to isolate the failure and more costly to replace the

Table 5.9 Comparison of BER Methods

	XOR	*FPGA*	*Digital Pin*	*Digitizer*
Throughput	Very high	High	Very high/high	Medium
Complexity	Very high	High	Medium	Low
Reusability	Low	Low	Medium	High
Hardware cost	Low	Medium	Very low	Low
Maintainability	Low	Low	Very high	High
Ease of development	Low	Low	Very high	High

failing part because either the complete load board must be replaced or a complex component must be replaced, which is costly and time-consuming. Lastly, it is easier to develop a solution using either a digital pin or a digitizer because the BER hardware and software are fully integrated into the test system; to develop with an XOR or FPGA is more difficult because they are not integrated into the test system.

5.22 BER Receiver Measurements

Below is a list of various BER receiver tests that are used for standards like Bluetooth and Digital Enhanced Cordless Telecommunications (DECT):

- Sensitivity BER;
- Carrier-to-interference (C/I) BER;
- Blocking BER;
- Intermodulation BER;
- Maximum input level BER.

All of these tests utilize BER as the criterion to determine receiver performance. All of these BER measurements are made by providing a digitally modulated signal to the device under test and comparing the digital bits produced at the output of the device with the original input bits.

5.22.1 Sensitivity BER Test

The sensitivity test measures the threshold power level when the receiver looses (drops) the incoming signal without any interfering signals. The sensitivity test is analogous to two people having a face-to-face conversation in which the first person continually lowers his voice until finally the second person can no longer hear or understand the first person. The Bluetooth specification states that with an input power to the receiver of –70 dBm, the receiver must be able to demodulate a DH1 signal with a BER result of less than 0.1%. Many suppliers want to characterize their devices to determine just how low of a signal power level the Bluetooth receiver can detect and demodulate correctly (i.e., how low can you whisper and still have the other person understand what you are saying). In this case, the input power level must be stepped and controlled accurately while a BER test is iterated. As an example, consider a generic Bluetooth radio modem that must pass the standard specification of –70 dBm. That is, the BER result must be less than 0.1% at an input power level of –70 dBm. But how many data bits should be used to perform this test? Is 1,000; 2,000; 10,000; or 100,000 the correct amount? The more data bits that are used for the test, the longer the test time will be and, ultimately, the higher the test cost. In addition, should the device be hopping frequencies during the BER test? In the final application, the device will be hoping frequencies while demodulating random data packets, but does the device really need to be tested in that manner. What should the BER data look like? That is, should it be random data? Chip manufacturers normally use a predefined algorithm of random data to approximate closely the types of data packets that the device will handle in the final application. Some of the answers to these questions are more straightforward than others, but in general, as long as the device adheres to the specifications as defined in the Bluetooth standard, then it qualifies to be sold as a Bluetooth-compliant device. The ability of the test equipment and test/product/application engineer plays a large role in determining how many of those questions are answered.

The remaining BER tests all involve the desired signal plus single or multiple interfering signals. The interfering signals can be either CW signals, modulated signals, or a combination of both, and all are used to approximate the real environment in which the wireless device must operate.

5.22.2 Carrier-to-Interference BER Tests

The C/I tests are performed with two signals, the modulated desired signal at a particular power level and another modulated signal as the interferer at another power level. The modulated interferer uses the same modulation format as the desired signal (Bluetooth, for example).

5.22.3 Cochannel Interference BER Tests

One of the most critical C/I tests is the cochannel test. The cochannel BER test for a Bluetooth device is performed by sending a cochannel Bluetooth-modulated signal simultaneously with the desired signal to the device under test and then measuring the receiver's BER. Both the desired signal and interfering signal are sent to the DUT on the same channel (at the same frequency). However, the data between the two channels is uncorrelated so that a Bluetooth device is still able to detect and demodulate the desired signal correctly as long as the power level of the cochannel interferer is not too high. The cochannel interfering data is usually taken to be pseudorandom bit sequence 9 (PRBS 9) data to approximate live traffic closely. This test determines if the Bluetooth receiver can operate while other Bluetooth devices are transmitting in the area on the same channel.

This test is similar to the situation where two people are trying to talk to you at the same time. If you concentrate on the first speaker, and the second speaker is not too loud, then it is fairly easy to understand the first speaker while completely ignoring the second speaker. However, if the second speaker gradually increases his voice, then at some point the second speaker's voice is simply too loud, and you can no longer understand the first speaker. Figure 5.22 is a graph showing the desired signal and the cochannel interferer for a Bluetooth device.

5.22.4 Adjacent Channel Interference BER Tests

It is also important to ensure that the wireless device can operate in the presence of nearby interferers. Adjacent channel BER tests are identical to the cochannel test except that the interferer is transmitted on one of the adjacent channels instead of on the same channel as the desired signal. For wireless devices, BER tests are often performed with interferers on the lower three and upper three channels. Figure 5.23 shows a plot of the three upper channel conditions used for a Bluetooth device. Notice that the power level of the interferer increases as the interfering signal gets further away from the desired signal (i.e., adjacent channel ± 3 has a higher

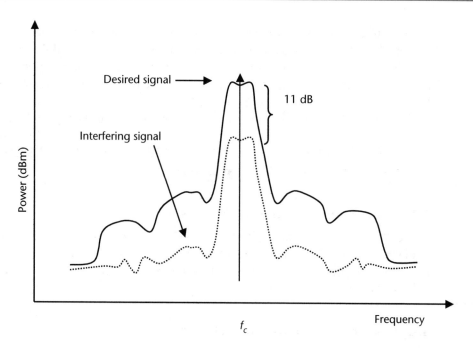

Figure 5.22 Cochannel BER test condition.

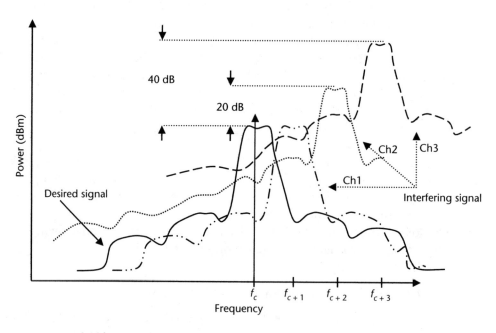

Figure 5.23 Adjacent channel BER test condition.

transmit power than does adjacent channel ± 2, which has a higher transmit power than does adjacent channel ± 1). The receiver's front-end filter has a specific mask characteristic that is designed to reject adjacent signals with predetermined power levels. The adjacent channel BER tests ensure the filter specifications.

5.22.5 Inband and Out-of-Band Blocking BER Tests

The blocking BER tests are designed to test both inband and out-of-band continuous wave interfering signals that can potentially interfere with the receiver's ability to detect the desired signal. Since the interfering signal is a single CW tone, it is referred to as a blocker because if the power level of the blocker is high enough, it can potentially block the receiver from detecting the desired signal. The front-end filter is designed to protect the receiver from inband and out-of-band blockers, but there is a limit to the amount of protection the filter can provide while allowing the desired signals to pass through unaffected. For Bluetooth, full compliance requires a continuous BER test while the blocking signal ranges from 30 MHz to 12.75 GHz in 1-MHz increments. Testing the complete range (30 MHz to 12.75 GHz) in 1-MHz steps would require 12,720 BER tests, and this is obviously impractical for cost-of-test reasons. Instead, BER blocking tests are performed at a few (usually fewer than five) critical points throughout the spectrum. These critical areas are worst-case points; if the device passes, it is guaranteed to pass the entire blocking specification.

5.22.6 Intermodulation Interference BER Tests

Intermodulation BER measures the receiver's nonlinear characteristics. Recall Section 4.20 on intermodulation and harmonics, where it was shown that for a non-linear device, intermodulation products are generated from the interaction of two CW tones. In the same manner, it is possible for two signals to interact with each other at the front end of the receiver to produce an intermodulation product that falls directly on top of the desired signal. If the power level of the intermodulation product is too high, it will completely inhibit the ability of the receiver to detect the desired signal, and all of the data will be lost. For a Bluetooth device, this test is performed by sending the desired Bluetooth-modulated signal in combination with two unwanted signals to generate third-, fourth-, and fifth-order intermodulation products. The BER is then measured to determine the receiver's performance in the presence of intermodulation distortion. It is similar to the traditional two-tone test that is used to test amplifiers with the additional complexity of three signals, two of which are modulated. The frequencies of the intermodulation interferers should be chosen such that

$$|F_2 - F_1| = n \tag{5.13}$$

where F_2 is a modulated signal using PRBS 9 data, F_1 is a single tone and n is equal to 3, 4, and 5 MHz, respectively. An $n = 3$ corresponds to the third-order intermodulation product; $n = 4$ generates the fourth-order intermodulation; $n = 5$ generates fifth order.

Figure 5.24 shows a plot of an intermodulation BER setup with an f_c of 2.450 GHz, $n = 3$, $F_1 = 2.47$ GHz, and $F_2 = 2.44$ GHz.

$$|F_2 - F_1| = n = |2.44 - 2.47| = 3 \tag{5.14}$$

where $n = 3$ is 3 MHz away from f_c (2.450 GHz).

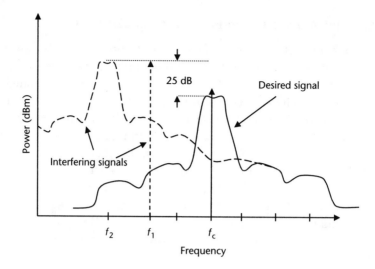

Figure 5.24 Intermodulation BER test condition.

It is too costly to test all combinations in production due to the extensive test time that would be required. Generally, chip manufacturers test their worst-case scenarios. Notice that there are three signals (desired, modulated interferer, CW interferer) on the plot. For this reason, this BER test is sometimes referred to as the three-tone test, because three distinct signals are required.

As an example, you can imagine a Bluetooth device operating in an area with other Bluetooth devices and perhaps a microwave oven as well. The CW tone generated by the microwave oven and an unwanted Bluetooth signal can interact inside the Bluetooth device to produce an intermodulation product that is at the same frequency (channel) as the desired signal. Because the intermodulation product lies on the same channel, the internal filter will pass the unwanted signal just as it passes the desired signal. The BER intermodulation test ensures that the intermodulation characteristics of the wireless device are within specifications.

The three-tone test requires that the test system has the ability to generate three distinct signals with two of the signals being modulated and the third signal being a CW tone. This means that ATE systems and test setups must have at least three separate signal generators to realize this test. As the Bluetooth/WLAN standards grow in popularity and other wireless devices become more prolific, variations in blocking and intermodulation tests that are more complex than the three-tone test will be needed. For example, a multiband Bluetooth/WLAN device may require BER testing with simultaneous interfering or blocking signals from both frequency bands to ensure the quality of the device. This multiple interfering BER test scenario would be the closest approximation to the real environment.

Imagine a typical office environment that utilizes both Bluetooth and WLAN networks. The multiband wireless SOC device must operate within such a noisy environment. Additionally, many of the employees will be making mobile telephone calls, adding more noise to the area. If the BER of the wireless device's receiver passes its specifications while being tested under similar conditions, then it will pass any subcondition testing scenario. This implies simultaneous interfering signals in the 802.11a and ISM band may be needed during testing. Since the frequency of 802.11a is at 5.6 GHz, second and third harmonics may be required, dictating that the signal path of the tester or test setup must operate over a 15-GHz band. This will present a significant challenge to ATE manufacturers because manufacturing cost-of-test systems becomes more expensive with increasing bandwidth requirements.

5.22.7 Maximum Input Power Level BER Test

The maximum input power level BER measures the receiver's ability to operate even when saturated or at the maximum power level. For Bluetooth, the specified input power level is –20 dBm. The maximum input BER test is similar to the sensitivity test, except that instead of measuring a low power-level signal, a high power-level signal is measured. The sensitivity and maximum input level tests in combination can be thought of as a dynamic range test for a Bluetooth receiver.

Care must be taken and various standards considered to ensure that a particular test setup will cover the many possible testing scenarios. For example, the maximum input BER test for the Bluetooth standard only requires a maximum input power level of –20 dBm, whereas the maximum input level for the DECT standard requires at least +13 dBm. To be able to test the DECT scenario, many test setups may require an amplifier to reach the higher power level of +13 dBm. If the test system does not have intrinsic high power capabilities, then a load board solution with a power amplifier is needed. The more components that are required on the load board, the more complex the calibration becomes, the more difficult tester-to-tester correlations become, and the more the reliability is reduced.

5.23 EVM Introduction[3]

Often in traditional RF and mixed-signal testing, measurements are made to infer the capability of a device within a communication system. Whether these tests include spectral tests, such as adjacent channel power, or single-frequency tone tests, such as signal-to-noise ratio (SNR), they do not test the actual ability of the device to perform properly in the system. A time-domain analysis of a modulated signal, such as the one used in the error vector magnitude measurement, can provide just this type of information.

5.23.1 I/Q Diagrams

Before proceeding into the details of error vector magnitude measurements, it is important to understand the concept of I/Q diagrams (also known as phasor diagrams). When analyzing wireless communication, the signal of interest is often a digitally modulated signal of radio frequency carrier. As such, viewing a simple voltage-over-time graph (e.g., a measurement from an oscilloscope) may not

---

3. This section was written by Ashish Desai.

provide much intuitive information regarding the quality of the signal. Therefore, the concept of an I/Q diagram is often used when describing digitally modulated signals.

An example will be the best way to describe this concept. First, define the data signals $I(t)$ and $Q(t)$, the carrier frequency w_c, and the transmitted signal $v(t)$. The following equation describes the relationship among these signals:

$$v(t) = I(t)\cos(w_c t) + Q(t)\sin(w_c t) \qquad (5.15)$$

While it is perfectly valid to analyze the signal $v(t)$, it is often easier to view only the signals $I(t)$ and $Q(t)$, with the understanding that the final signal is actually modulated at the carrier frequency w_c. The I/Q diagram does just this by plotting the signal $Q(t)$ versus $I(t)$. The $I(t)$ is referred to as the in-phase component (traditionally, the cosine function is taken as the reference for phase), while the $Q(t)$ is referred to as the quadrature component (the sine function is orthogonal to the cosine function).

Digital modulation techniques typically assign bit values either to certain points within the I/Q diagram (also known as a constellation) or to a particular transition between sets of these points. Therefore, it is much easier for one to determine the data that is being transmitted by viewing an I/Q diagram rather than a voltage-over-time graph. Figure 5.25 shows an example of an I/Q constellation diagram for the modulation format quadrature phase shift keying (QPSK). This format includes four points in the I/Q plane, each with equal amplitude (distance from the origin) but with four different phase shifts (angle between I axis and the vector connecting the origin and the constellation point).

One should also note that this method of viewing modulated signals could also be used with devices that transmit at an intermediate frequency (IF) or with baseband signals equivalently.

5.23.2 Definition of Error Vector Magnitude

A particular type of measurement that recently has become more popular in measuring the signal quality of digitally modulated signals is the error vector magnitude (EVM) measurement. To perform this measurement, a modulated signal is captured

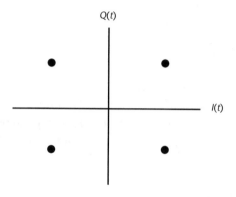

Figure 5.25 Example I/Q constellation of a QPSK modulated signal.

in the time domain and then plotted with an I/Q diagram. It is then compared to a reference signal (created using the same data), and the difference between the measured and reference signal is referred to as the error vector. The EVM is defined as the ratio of the magnitude of the error vector and the magnitude of the reference vector (a vector that begins at the origin of the I/Q constellation and ends at the reference point). Mostly, this ratio is described in a percentage form. An example of the measured, reference, and error vectors is shown in Figure 5.26. Although this type of measurement is referred to as EVM, often other figures of merit are also computed when making this measurement. Phase error is defined as the difference (expressed in degrees) in the phase of the measured vector from the reference vector. Magnitude error is defined as the ratio (described in decibels) of the measured vector and the reference vector. Frequency error is defined as the difference between the expected carrier frequency and the measured frequency (determined from the speed with which the constellation is rotating).

5.23.3 Making the Measurement

While the definition of the EVM measurement seems straightforward, often this can be a challenging figure of merit to compute. Traditionally, power measurements have been the staple of analysis of RF modulation, and therefore the equipment associated with the testing of RF components may not have the capability to perform time-domain analysis. For example, many spectrum analyzers contain diodes to compute average power, which means that the voltage data as a function of time cannot be extracted from the equipment. A typical instrument with the proper capability to perform time-domain analysis of an RF-modulated signal will have a block diagram similar to the one shown in Figure 5.27(a). The input will have a mixer with a system local oscillator to downconvert the RF signal to an intermediate frequency (IF) signal. Then, the IF signal will be digitally sampled by an analog-to-digital converter (ADC), at which point the time-domain trace will be passed to a computational unit [e.g., a digital signal processor (DSP), a separate workstation]. This computational unit will then attempt to analyze the signal by comparing it to the reference. While it is not necessary for the test equipment to include each of the

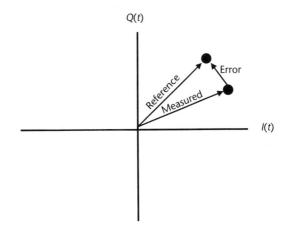

Figure 5.26 Example of measured, reference, and error vectors.

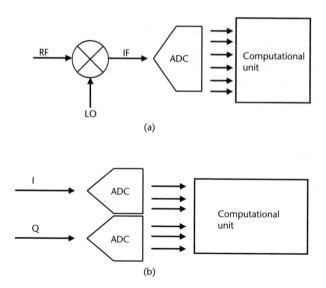

Figure 5.27 Typical EVM tester block diagrams: (a) RF EVM tester block diagram, and (b) baseband EVM tester block diagram.

components listed here [e.g., the device could directly downconvert RF to I/Q rather than to an IF or may have baseband outputs as in Figure 5.27(b)], it is required that the capability to analyze time-domain data be included. Of course, in assessing the test equipment, one must also take into account the bandwidth of the signal that needs to be measured. While some of the earliest forms of digital modulation only require measurement bandwidths of a few hundred kilohertz, the latest generation formats such as those used in wireless LAN can use upwards of 20-MHz bandwidths. Because these measurements are made in the time domain, it is important to realize that the ADCs used in the test equipment must have the capability to capture the entire bandwidth in one measurement (whereas power measurements may be able sweep the frequency spectrum by making multiple measurements).

At first glance, one may feel that given a time-domain signal and a reference signal, the computation of the EVM is very straightforward. This is very true; however, the computation of the reference signal can be very tricky and not for the faint of heart. First, remember that the reference signal must be created using the data to be transmitted. If prior knowledge of this data is not available, any algorithm to create the reference signal must first try to estimate the transmitted data. Therefore, it is possible to introduce error into the measurement itself by incorrectly estimating the reference data. Second, assuming the transmitted data is known or can be determined, the reference data and the measured data must be synchronized in time to compare the signal quality truly. (Note: Both traces are typically sampled higher than the symbol rate so that the symbol clock can be recovered.) Otherwise, the EVM measurement can be a very misleading way of determining the actual behavior in a communications system. For example, a perfect signal that has zero bit errors may have a high EVM if the reference and measured traces are skewed slightly in time (i.e., the symbol clock was incorrectly recovered). Finally, assuming that both the data estimation and the timing synchronization issues have been resolved, one must also be careful to make sure frequency synchronization is achieved. If the frequency that is used to make the measurement and the actual frequency that is transmitted are not equal, the entire modulated signal will rotate in the I/Q plane at a rate that is determined by the difference in these frequencies. While this is certainly an important error to detect, one should keep in mind that often communications systems will correct for frequency reference differences between the transmitter and receiver. Therefore, the EVM measurement should first determine this frequency difference and then adjust the reference signal to prevent this type of error from artificially degrading the measure of signal quality.

Because of these complexities, many pieces of test equipment include algorithms to measure EVM for various modulation formats, rather than simply providing an end user with an I/Q trace to analyze themselves.

For an ATE system, often the transmitted signal is provided to a device and the output is also measured by the system. In this case, the complexities associated with creating the reference signal can be greatly reduced, as the input data is known and the frequency error should be minimal between the source and receiver of the test system. This leaves only the propagation delay through the device to align the symbol times between the measured and reference signals. This delay can be computed through techniques similar to using an eye diagram to determine symbol detection in digital systems.

5.23.4 Related Signal Quality Measurements

While the definition for error vector magnitude given above is complete, many times other related time-domain measurements are grouped in the same category as EVM. This is especially true for modulation formats in which the transmitted data must follow a specific pattern, rather than be completely randomized. For example, with 3G CDMA modulation, there are many algorithms to measure various aspects of the signal in relation to the code domain that are often considered part of the EVM measurement. These measurements take advantage of the fact that only certain sequences of data are allowed, and therefore, the estimation of the reference signal must take into account a more complicated algorithm for detecting symbols than simply finding the symbol that has the smallest distance in the I/Q plane. Another example in which there may be other interpretations of the EVM measurement is multiple carrier channel modulation formats, such as OFDM. For the 802.11A standard of OFDM, there are 52 separate carriers, each with its own modulated signal (e.g., BPSK, QPSK, QAM64). In this case, one must be careful when describing the EVM of this signal for there is an EVM measurement associated with each carrier (as well as an I/Q diagram for each), as opposed to the case where only one carrier exists. Often in this case, an average EVM and a peak EVM are reported for the entire group of carriers that are measured. While many pieces of test equipment can report these other related signal-quality measurements, understanding the basic EVM measurement can often provide enough insight for an average user. This is especially true in a production environment where the focus is on pass/fail criteria, rather than detailed characterization of various aspects of a DUT (for more details on production testing using EVM, see Section 5.23.6).

5.23.5 Comparison of EVM with More Traditional Methods of Testing

Traditionally, RF devices are tested using various power measurements in the frequency domain (modulated carrier power, adjacent channel power, third-order intercept, and so forth) with little to no focus on the time-domain signal. While power measurements can quantify the performance of specifications such as gain and noise figure, these measurements may not accurately reflect the performance of a device within a communication system. To this end, the use of the EVM measurement can advise the user more concretely on the integrity of a signal through the component under test.

Consequently, it may be possible also to eliminate some of these traditional tests that are used to infer performance by the more direct measure in the time domain. Because the various analyses in the frequency domain each often focus on a particular frequency range (e.g., carrier power focuses on main channel, ACP on adjacent, TOI on an intermodulated frequency), many frequency-domain measurements are often made. If the time-domain signal can be compared to an ideal signal, then one simple measurement is sufficient to determine the pass/fail decision. Of course, to compare this to measurements with out of band noise truly (ACP, TOI), one would have to place an interfering signal at these same frequencies (e.g., source both the signal of interest and an adjacent channel into the DUT) and use the test equipment to filter out the interfering signal.

While the correlation of the values returned by the EVM measurement and traditional tests may not be entirely intuitive, often the I/Q diagram can be very helpful in determining errors. For example, an amplifier device that may have problems with its leveling circuitry will have a constellation that has varying amplitude (and only amplitude), as shown in Figure 5.28(a). If a baseband device has an amplitude imbalance between the I and Q channels, the constellation will become stretched in one direction, as shown in Figure 5.28(b). If there is a frequency offset between the receiver and transmitter, the constellation will rotate in the I/Q plane, as shown in Figure 5.28(c). Simple spreading of the constellation points around the ideal point can occur because of problems with thermal noise or intersymbol interference (caused by nonideal frequency response through the DUT), as shown in Figure 5.28(d).

5.23.6 Should EVM Be Used for Production Testing?

In ATE, the focus is always to reduce the cost of testing a device. To this end, typical cost reduction includes either lowering the capital costs of the test equipment or reducing test time. Using the EVM measurement may achieve this reduction in test time by replacing some of the more traditional measurements of RF devices.

However, one must consider the trade-offs in making an EVM measurement. For example, if the test equipment one uses does not perform the measurement with a computational unit that resides on-board with the ADC, then there will be transfer time associated with moving the data to the computational unit (typically a workstation). As the bandwidth of the modulated signals increases, the required sample rate of the digitizer also increases, leading to a higher number of samples that need to be transferred. Additionally, the repeatability of the EVM measurement increases as more symbols are compared, again leading to more samples that need to be

5.23 EVM Introduction

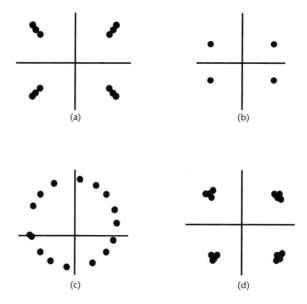

Figure 5.28 Various QPSK modulated signals with impairments: (a) amplitude error (time varying), (b) amplitude balance error, (c) frequency error, and (d) ISI, or thermal noise, error.

captured. Therefore, for large bandwidth signals with a large number of symbols, the measurement time for this type of analysis may become prohibitively large. Of course, the impact of these parameters will vary, depending on the test equipment (transfer times, workstation speeds, digitizer features, and so forth).

Often, signal quality measurements such as EVM can be an excellent tool during development of a design, but thus far they have not been used heavily in production, probably due both to the fact that traditional methods are more accepted and have not been closely correlated to EVM, as well as to the fact that the test-time improvements are dependent on the many factors mentioned in the previous paragraph. Unfortunately, one still needs to determine on an application basis whether this time-domain analysis is both sufficient and useful in testing a device. While a definitive answer has not been provided, hopefully the correct considerations have been included to help one make an educated decision as to whether EVM measurements can be used in a particular application.

References

[1] Agilent Technologies, "Performing Bluetooth RF Measurements Today," Application note 1333, March 9, 2001.
[2] Robinson A., "On Your Marks for Testing Bluetooth," *Test & Measurement* Europe/ June–July 2000.
[3] Bluetooth RF Test Specification, revision 1.1, February 22, 2001.
[4] Bluetooth Special Interest Group, at www.bluetooth.com, June 2003.
[5] Bray, J., and Sturman, Charles F., *Bluetooth Connect without Cables*, Upper Saddle River, NJ: Prentice Hall, 2001, pp. 17–24.
[6] Blanchard, A., *Phased-Locked Loops*, Hoboken, NJ: John Wiley & Sons, 1976.

CHAPTER 6
Fundamentals of Analog and Mixed-Signal Testing

Edwin Lowery III

6.1 Introduction

The trend of RF microchip development for the past decade has been that of higher and higher levels of integration. Each new generation of chips seems to be more and more complex, containing more and more elements of the radio and fewer total parts in the end-product bill of materials. One of the interesting developments in modern SOC test methodologies is that as integration of the RF transceiver becomes more and more complete, there is a collision of RF and mixed-signal methodologies to test these devices. Understanding RF alone is no longer sufficient in order to test a device with numerous baseband components. Likewise, engineers with a mixed-signal background wishing to analyze a signal that has been upconverted to the gigahertz range will need to use a variety of RF techniques.

Each discipline, RF and mixed signal, has a series of conventions describe what it does. Interestingly enough, many of these concepts are extremely similar to one another, explaining the same phenomenon from different points of view. RF engineers tend to concentrate on energy flow and how this energy is reflected and absorbed by a load. Also, RF engineers spend a lot of time thinking about signals in the frequency and quadrature domains. Mixed-signal engineers tend to concentrate on digitized representations of signals in the frequency and time domains. They are primarily concerned with how to sample a signal while adding minimum distortion and how to re-create a signal with minimum samples.

This chapter will attempt to bridge the gap between the two fields, concentrating on the mixed-signal realm. It will explain the basics of practical mixed-signal test engineering and provide a map of how to translate back and forth between common RF and mixed-signal concepts.

6.2 Sampling Basics and Conventions

Since the language of the mixed-signal engineer seems to differ so much from that of RF engineers, this section will attempt to define commonly used terms for AWGs and digitizers. More information and details can be found in the List of Acronyms and Abbreviations at the end of this book.

6.2.1 DC Offsets and Peak-to-Peak Input Voltages

Special consideration should be given to the concepts of peak-to-peak voltage and dc offset (Figure 6.1). In the RF realm, practically all signals analyzed have little or no dc component. If there is a dc component, most measurement equipment will capacitively decouple it so that the engineer can focus on the signal of interest with maximum dynamic range.

For mixed-signal engineering, the dc offset and the peak-to-peak input voltage are paramount. Modern digitizers and AWGs are specified with a variety of output voltage ranges, as well as dc offsets. In order to introduce a minimum amount of distortion into a signal, the more that is known about the signal the better.

Assume that a digitizer that is available in an ATE system has an input voltage range of ±2V, ±1V, ±250 mV and ±250 mV. Take the example of a signal with a dc offset of 200 mV and a peak-to-peak voltage of 230 mVs. It would be tempting to use the ±500-mV range and hope for the best. Since modern day ATE digitizers have the ability to account for a dc offset, it would be much better to use a 230-mV dc offset and a range of ±250 mV. The best way to think about this is in terms of the size of the least significant bit (LSB). The way to minimize any sampling error and maximize dynamic range is to pick the smallest LSB possible.

$$LSB = \frac{MaxRange}{2^{bits} - 1} \qquad (6.1)$$

For a 12-bit digitizer with the input range set to ±250 mV, the step size would be 244 μV. The step size for the ±250-mV range will be 122 μV. If we assume that the error for each sample will be ±1/2 LSB, the error per measurement will be less than half for the smaller range than for the larger.

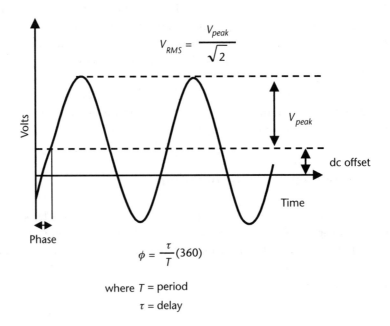

Figure 6.1 DC offset and peak-to-peak voltages.

6.3 The Fourier Transform and the FFT

Once the signal is digitized, analysis can begin. Since so much of mixed-signal engineering occurs in the frequency domain, it will greatly help our understanding to begin by looking at the mathematics behind converting a signal to the frequency domain.

6.3.1 The Fourier Series

One of the cornerstones of frequency analysis is the concept of the Fourier series. The Fourier series shows that virtually any periodic waveform can be represented as a sum of discrete sinusoids. Of course, there are rigorous conditions that must be met first: The waveform must be periodic, it must be integratable over any period (no infinite discontinuities), and it must be valid for all values of time [1].

$$x(t) = a_0 + \sum_{n=1}^{\infty} \left(a_n \cos(n\omega_0 t) + b_n \sin(n\omega_0 t) \right) \tag{6.2}$$

This is a great concept. It allows us to represent any periodic signal by a series of discrete sinusoids. But what happens if the signal is not periodic? Almost all real-world signals have a start time and a stop time, and they are therefore not valid for all time. We need a different theory that covers this phenomenon.

6.3.2 The Fourier Transform

Fourier understood that almost all signals that we are working with have a finite duration, and he went further to describe a time-frequency relationship for nonperiodic signals. This relationship is called the Fourier integral or Fourier transform. Of course, the time-domain signal still has to meet the minimum requirements of being finite and integratable over the interval for which it is defined [2].

$$X(f) = \int_{-\infty}^{\infty} x(t) e^{-j2\pi f t} \, dt \tag{6.3}$$

Also, Fourier defined a way to translate a continuous signal from the frequency domain to the time domain. This equation is called the inverse Fourier transform (6.4) [2].

$$x(t) = \int_{-\infty}^{\infty} X(f) e^{j2\pi f t} \, df \tag{6.4}$$

This chapter will not spend a lot of time going over all of the rigorous mathematics, but using these equations, several common transformations have been derived. Figure 6.2 shows some of the most commonly used transformations.

The cosine transformation shows that taking a signal from the time domain into the frequency domain gives you two frequency components, one at 1/T, the other at –1/T. Adding the two amplitudes together in the frequency domain gives us the amplitude in the time domain. For now, the phase relationship is ignored. This is

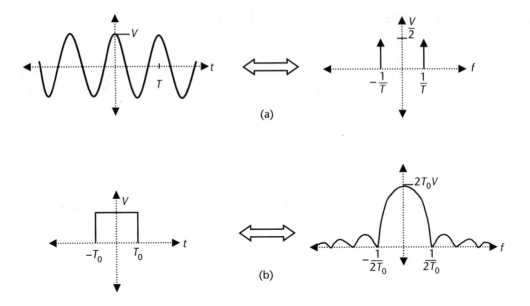

Figure 6.2 Commonly used Fourier transforms: (a) cosine transformation, and (b) rect pulse transformation.

useful to show that a time-domain sinusoid (sine or cosine) is represented as a single tone with both positive and negative components in the frequency domain.

Figure 6.2(b) shows another very common transformation, the rectangular pulse transformation. Starting with a rectangular pulse in time centered on $t = 0$, we note that its height is V, and its width is $2T_0$. This translates to a sinc pulse (its absolute value is shown here), which is centered on 0 in the frequency domain. Its amplitude is $2T_0V$, and its first zero crossings occur at $\pm 1/2T_0$. The rectangular (rect) pulse is used to represent a sample in time of another signal. Basically, any signal that we are interested has a start and a stop. The rect pulse allows us to take a continuous signal and mathematically chop it down to a specific time interval. Using concepts that will be introduced shortly, this windowing technique of using the rectangular shape as a reference will be explored in detail.

Another extremely useful concept in Fourier analysis is the concept of the impulse signal, or $\delta(t)$. An impulse is merely an extremely short burst of energy in either domain. It occupies very little time, and its height is usually 1. It turns out that this is a great way to represent a single frequency component, or even a sample in the time domain. Figure 6.3 shows some common relationships that impulse signals have between the time and frequency domains [2].

The first relationship in Figure 6.3(a) shows that a dc signal, which does not change in time, has a single frequency component at zero. Also an impulse signal in the time domain contains a continuous, or infinite, frequency response. Another way of stating this is, signals with an infinitely small time response have an infinitely large frequency response.

The last and perhaps most important relationship is the impulse train transformation. A series of impulses or comb function in the time domain transforms into a series of impulses or another comb function in the frequency domain. Sampling in

6.3 The Fourier Transform and the FFT

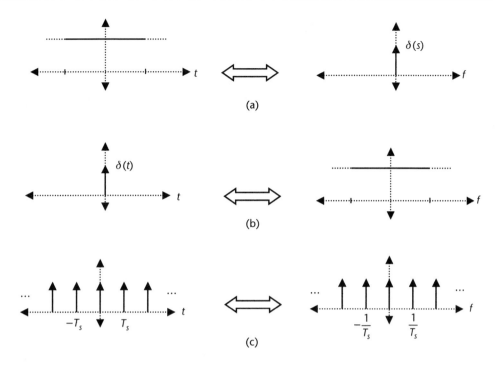

Figure 6.3 Commonly used impulse transformations: (a) dc transformation, (b) impulse transformation, and (c) impulse train transformation.

the time domain leads to samples in the frequency domain. If the samples are spaced T seconds apart, then samples in the frequency domain will appear to be spaced by $1/T$ in the frequency domain. This concept is explored further in Section 6.5.

6.3.3 The Discrete Fourier Transform

The Fourier transform is quite a useful stepping-stone, but it makes an assumption that we have access to a continuous signal in the time domain and that the result is another continuous signal in the frequency domain. This is useful in helping us to understand the relationship between the time domain and the frequency domain; however, some tools are needed to help us work with a sampled version of a continuous waveform. To transform a discrete sampled waveform of N samples from the time domain $x(n)$ to the discrete frequency domain $X_d(k)$ we use the discrete Fourier transform (DFT) [2].

$$X_d(k) = \sum_{n=0}^{N-1} x(n) e^{-\frac{j2\pi kn}{N}} \qquad (6.5)$$

Of course, there is a way to take this discrete frequency domain and translate it back into the time domain, and this is accomplished by using the inverse DFT (6.6) [2].

$$x(n) = \frac{1}{N} \sum_{k=0}^{N-1} X_d(k) e^{\frac{j2\pi kn}{N}} \qquad (6.6)$$

If you go through the mathematics involved in actually calculating the DFT, it gets quite intensive. In fact, the DFT was never really used in the field of mathematics, except in scientific study, and then only with the use of computers. The computations needed make hand analysis impractical.

Take as an example a 1,024-point sampled time-domain signal. In order to translate this into the frequency domain, N^2 calculations are needed (N in this case = 1,024.) This means that 1,048,576 calculations must be performed before the transformation is complete. This is a lot of time for hand calculations. The DFT has also been attempted in real-time computing systems, but this many calculations for very time-dependent results is too expensive, even for today's fast computer systems. Another methodology is needed.

6.3.4 The Fast Fourier Transform

Thankfully, looking carefully at symmetries and the built-in relationships that the DFT terms have with each other, the fast Fourier transform (FFT) was developed. It drastically decreased the processing steps required to translate discrete signals from one domain to another. It provides exactly the same result as the DFT, although it takes advantage of the frequency-domain symmetry and reuses of some seed terms to provide the positive frequency spectrum with much fewer calculations. The FFT requires only $Nln(N)$ steps to perform its conversion as long as N is an integer power of two. For our example of 1,024 points, this means there are only 7,098 operations. The DFT by comparison takes 147 times longer. Needless to say, the FFT is predominantly what is used today, and that is why most FFTs are calculated in sample sizes of 2^n.

6.4 Time-Domain and Frequency-Domain Description and Dependencies

Most of the mixed-signal engineer's time is spent going back and forth between the time domain and the frequency domain. In order to dig deeper into these relationships, some new concepts need to be introduced.

6.4.1 Negative Frequency

The first concept that was sort of glossed over is that of negative frequency. When we start to discuss the Fourier transforms of signals from the time domain to the frequency domain, there is always a positive and a negative frequency component. What does this mean? The short answer is that we typically ignore the negative component, but in order to really understand what is happening to signals in the frequency domain, we have to understand that negative frequency components exist.

When we perform a rigorous analysis of the mathematics of the Fourier transform, we see that in order to transform a signal from the time domain into the frequency domain, a few conditions must be met. For convention, we represent a continuous function in the time domain as $x(t)$ and its Fourier transform in the frequency domain as $X(f)$. First, the signal must be continuous (i.e., $x(t)$ must have a real value for any value of t). Next there must be a finite integral over any interval of the signal, or stated differently, any discontinuities must be finite.

Once these conditions are met and we mathematically translate a signal from the time domain into the frequency domain, we find that if $x(t)$ is real, the magnitude for the $X(f)$ positive axis has a mirror image of itself projected onto the negative frequency axis. Also, the magnitude of the frequency-domain component is equal to exactly half of the magnitude of the signal present at any point. This means that the total value of the frequency component at frequency f is equal to $X(f) + X(-f)$. It is important to note that symmetry is not maintained if $x(t)$ is complex and contains real and imaginary components.

Incidentally, if $x(t)$ represents a real continuous time-domain function, and $X(f)$ represents its Fourier transform, $\Phi(f)$ represents the transform's phase response over frequency. For the case of $x(t)$ being only real, the positive and negative phase axis contains an inverted mirror image of the results. In other words $\Phi(f)$ is equal to $-\Phi(-f)$ for real $x(t)$.

6.4.2 Convolution

The convolution is defined in the following equation. It is also abbreviated as $a(x) \otimes b(x)$. (Note: The $\otimes$ sign denotes convolution and should not be confused with the multiplication sign.)

$$\int_{-\infty}^{\infty} a(u)b(x-u)du \tag{6.7}$$

The convolution is best explained by a graphical example, as shown in Figure 6.4. The idea is that the first signal is flipped along its horizontal axis, and then slid along the top of the other signal on its horizontal axis. As it moves from left to right, a sum, or integral, of the areas that touch each other is performed. The resultant plot shows that this produces a kind of "smear" of the original signal. The use of the convolution is key to understanding the concept of aliasing. For this discussion the exact mathematics are not as important as the graphical concept.

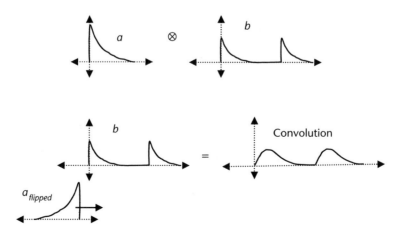

Figure 6.4 Convolution example.

6.4.3 Frequency- and Time-Domain Transformations

Building a complete understanding for what happens in the process of translating a signal from the time domain to the frequency domain and vice versa has been the subject of many texts over the years. For the purposes of brevity and overview, here are some of the most important concepts to keep in mind (Figure 6.5).

The first relationship is that of linearity. The linearity relationship states that if two signals are added together in the time domain, the results are the same as adding together each of their individual Fourier transforms. This is important because if we understand how to break down a time-domain signal into its individual components, we can then find the Fourier transform of each component and combine their respective results in the frequency domain. The results will be the Fourier transform of the original.

The next two concepts are somewhat intertwined. These are the concepts of time compression and frequency expansion. Simply put, compressing a signal in the time domain expands it in the frequency domain, while reducing its magnitude. Recall that in Figure 6.3(b) the impulse is the most compressed signal that can exist in time, and it contains an infinite frequency response. Stretching a signal out in the time domain compresses it in the frequency domain. This concept is also demonstrated by Figure 6.3(a). The signal, which is completely compressed in the frequency domain, is stretched out entirely in the time domain.

To demonstrate this property of time compression and frequency expansion more thoroughly, a simulation was performed in Matlab, which found the Fourier transform of a rectangular pulse. Figure 6.6 shows the results of this analysis. Figure 6.6(a) shows a time-domain pulse that is 5 samples wide and 1,000 samples long. Recall from our previous discussion of Fourier transforms that if a square wave has a width of T and a height of V, its Fourier transform has a height of $2TV$, and the zero crossings will be at $\pm 1/(2T)$.) In the case of the discrete Fourier transform, this is only slightly modified. For our example, note that $V = 1$ and $2T = 5$. Therefore, the height of the transform is 5. Since there are 1,000 samples, the first zero crossings are at $\pm N/(2T)$, where N stands for the number of samples. This puts the zero crossings at ± 200 for the upper left-hand trace. Now, take a look at the upper right-hand plot, Figure 6.6(b). Notice that by setting the width $2T$ at 51, since

Time domain	Frequency domain	Concept
$x(t) + y(t)$ ⟷	$X(f) + Y(f)$	Linearity
$x(kt)$ ⟷	$\dfrac{X(\frac{f}{k})}{\|k\|}$	Time compression yields frequency expansion and vice versa
$x(\frac{t}{k})\|k\|$ ⟷	$X(kf)$	Time expansion yields frequency compression and vice versa
$x(t-T)$ ⟷	$X(f)e^{-j2\pi fT}$	Time domain shift yields phase shift in frequency domain
$x(t)e^{j2\pi tF}$ ⟷	$X(f-F)$	Frequency shift causes time-domain modulation

Figure 6.5 Relationships between the time domain and the frequency domain.

6.4 Time-Domain and Frequency-Domain Description and Dependencies

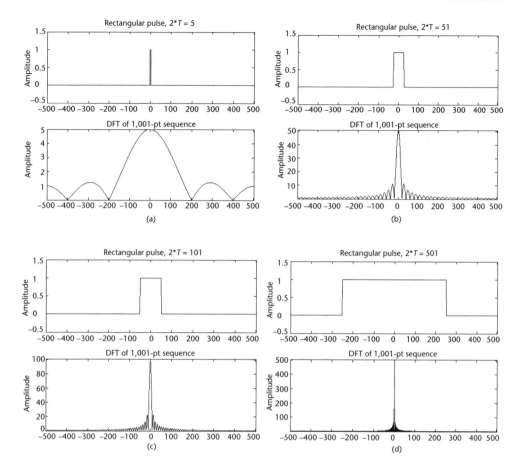

Figure 6.6 Examples of time and frequency relationships: (a) $2T = 5$, (b) $2T = 51$, (c) $2T = 101$, and (d) $2T = 501$.

the pulse is wider in the time domain, it becomes narrower in the frequency domain. A width of 51 was chosen to maintain an even number of samples on either side of the horizontal axis. Also, since $2T$ has increased, the maximum value for the FFT is also increased to 51. The first zero crossings now occur at 1,000/51 or 19.6. This shows that, indeed, expanding a signal in the time domain causes amplitude increase in the frequency domain, along with compression along the frequency axis.

Figure 6.6(c, d) further illustrates what happens as the pulse width continues to increase.

In order to shift a signal in phase, using complex mathematics we multiply by $e^{-j2\pi f}$ in the frequency domain. This is the next property in Figure 6.5. It states that a shift in the time-domain signal causes a phase shift in the frequency domain. This is useful because once we know the Fourier transform of the shape of a waveform in the time domain, applying a shift in time to that signal will not affect the magnitude of its FFT; it will only affect its phase. Finally, a shift in the frequency domain causes modulation in the time domain. Multiplying a signal by $e^{j2\pi tf}$ in the time domain means multiplying by a complex frequency. For real time-domain signals, it means multiplication by a sinusoid or tone. Another way to state this is that shifting a signal in the frequency domain is achieved by multiplying by a sinusoid in the time domain.

6.5 Nyquist Sampling Theory

RF engineers tend to connect their test equipment to the device inputs and outputs and immediately begin to look at the signals in the frequency and phase domain without considering how exactly they are digitized and converted. For mixed-signal engineers this process is essential and visited and revisited quite often.

The core scientific principle used in sampling a waveform is called Shannon's theorem. It states that if a function $x(t)$ exists with a frequency no higher than F_{max}, it can be completely reconstructed from a sampled version $x[n] = x(nT_s)$ if the samples are taken at a rate $F_s = 1/T_s$ such that $F_s > 2F_{max}$ [3]. In other words, in order to reconstruct the analog signal completely from the digitized version of the analog signal, we must sample at a frequency at least twice as high as the maximum frequency of the analog signal. That minimum sampling frequency is called the Nyquist sampling rate.

The process of what happens to a sampled signal is shown graphically in Figure 6.7. The first line shows an idealized analog signal $x(t)$, which is converted to the frequency domain. Note that the signal contains both a positive and negative frequency representation. Since the time-domain signal is real, its negative frequency component is a mirror image of the positive one. Also, note that the frequency spectrum is bounded by F_{max} and $-F_{max}$.

By sampling $x(t)$ we are basically multiplying an impulse train or comb function by a time-domain function. Recall that multiplication in the time domain

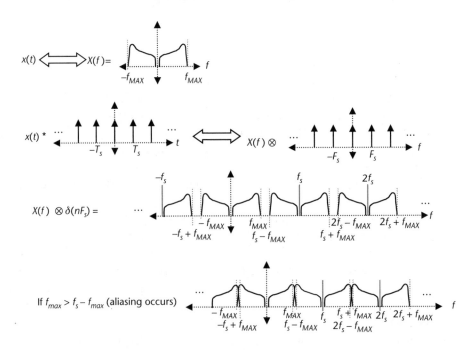

Figure 6.7 Sampling and aliasing.

corresponds to convolution in the frequency domain. In fact, we end up convolving the frequency representation of $X(f)$ by another comb function in the frequency domain. What we end up with is infinite copies of both the negative and positive frequency components of our time-domain signal.

Each copy of the spectrum is centered around integer multiples of the sampling frequency F_s so that we have copies around $F_s - F_s, 2F_s, 3F_s$, and so on.

Take a close look at the border between the first frequency representation and the second one. Notice that the highest frequency content of the original spectrum is at F_{max}. Also notice that the lowest frequency content of the next spectral copy is at $2F_s - F_{max}$. If F_s is too small, such that $2F_s - F_{max}$ is less than F_{max}, then aliasing occurs. The last line of Figure 6.7 shows the concept of what occurs in an aliased signal. Basically, higher-order copies of the spectrum get folded back into our base spectrum, which corrupts the signal. Aliasing should be avoided whenever possible as it degrades the fidelity of our sampled signal.

In the real world, there is no such thing as a signal that is completely bounded in frequency. There is always some spectral leakage outside of the band such that when we assign a maximum signal frequency, it really means the maximum frequency of interest. Looking at any spectral plot, noise is always present across the entire band. How is aliasing handled in the real world? Figure 6.8 shows a typical frequency response of a continuous signal. The vertical dimension is decibels and horizontal is frequency. This shows that while we may only be interested in the signal up to a specific frequency, it certainly will contain energy and content above that F_{max}.

Looking in the sampled frequency domain, aliasing will certainly occur that corrupts the signal of interest. What is typically done is that the signal is first filtered by a lowpass antialiasing filter before it is digitized. This minimizes the frequency content above F_{max}. Adding a lowpass analog antialiasing filter at the front end of test equipment greatly reduces the amount of aliasing that occurs, but of course some finite amount of error still exists due to the folding of the spectral content in the cutoff band of the filter. An ideal filter would completely remove all signals above f_{max}; however, real-world filters will only attenuate and distort signals above f_{max}. The idea here is that the filter will help reduce error, but there is no way to remove it

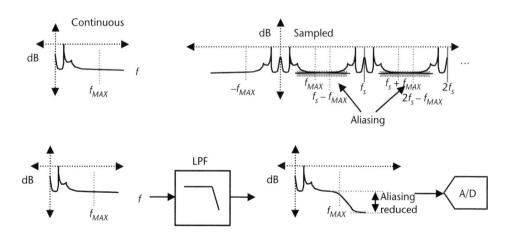

Figure 6.8 The antialiasing filter.

completely. This is the primary reason why mixed-signal engineers are so interested in the noise relative to the desired signal.

6.6 Dynamic Measurements

6.6.1 Coherent Sampling and Windowing

While the Fourier transform of a sinusoid may be a single tone in the frequency domain, exact representations of this using the FFT are rarely achieved in the real world. The problem lies in knowing exactly what the frequency is of the signal we are trying to sample and calculating the number of points needed. To be able to capture a sinusoid and create its FFT in the frequency domain with minimal noise, we need to know that the phase of the signal where digitization begins is exactly the same as the phase at the end of the sample. This is called coherent sampling.

No matter how a waveform is sampled, it is still just a sample of the original signal. Even though we don't realize it, we are applying a "window" that we use to look at the original signal. By capturing a signal coherently, we are ensured that the window that we use to look at the signal does not corrupt the original signal.

In modern ATE systems it is true that we often know what the frequency is that we are measuring, because we are the one supplying it. In this case coherence is easy to maintain; we simply make sure that we capture an integer number of cycles of the desired waveform. To do this, simply set up the ATE system such that the following relationships are met [3]:

$$\frac{F_s}{N} = \frac{F_t}{M} \tag{6.8}$$

where

- F_s stands for the sampling frequency or frequency of the digitizer.
- N stands for the total samples taken.
- F_t stands for the frequency of the signal we are measuring.
- M stands for the number of cycles of the measured signal that we capture.

What happens in the case where phase continuity is not guaranteed and the window does affect the signal? What if it is not possible to maintain the relationship in (6.8)? How do we model this?

Think of a captured time-domain waveform as a signal that is multiplied by a rectangular pulse. By multiplying in the time domain with a rectangular pulse, we get smearing in the frequency domain. The amount of smearing depends on the phase discontinuity of the signal in question.

Figure 6.9(a) shows an example of a sampled sinusoid with a very large phase discontinuity. Notice that there are just over five cycles captured, and while the phase of the signal starts at zero with the waveform rising, it also ends at zero with the waveform falling. This is equivalent to an almost 180° phase shift from end to beginning. The frequency plot on the right, in Figure 6.9(b), shows the smearing

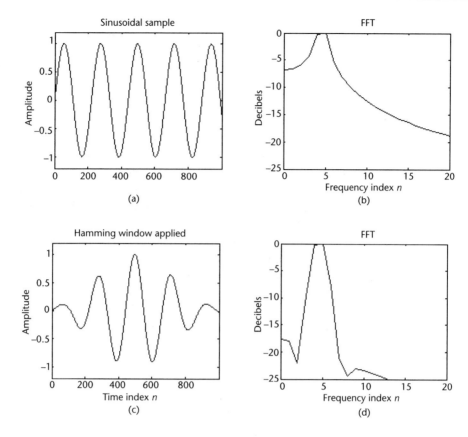

Figure 6.9 Noncoherent sampling and Hamming window: (a) noncoherent time domain, (b) noncoherent frequency domain, (c) windowed time domain, and (d) windowed frequency domain.

effect of this kind of phase discontinuity. The frequency-domain samples are referred to as bins. The term *frequency bin* refers to the sampling rate divided by the number of samples. The frequency-domain plot has been scaled to show only the first 20 samples, or frequency bins. Since there are roughly five cycles captured, the peak of the FFT is in bin 5.

Digital signal processing has many different methods to handle this kind of situation, but one approach is first applying a specific window to the time-domain signal. Instead of using a rectangular window, which is what we are doing in the upper left-hand corner, we try and de-emphasize the phase discontinuity at the ends by multiplying the time-domain signal with a more gradual envelope that is near zero at the beginning and end of the time domain. This gradual shift to zero has a minimum effect on the amplitude of the capture signal, but a major effect on reducing the smear effect. There are many different types of windows that can be used. The most common are Hamming, Hanning, Blackman, and Kaiser. There are entire texts dedicated to generating your own windows and using existing ones, but most of the common windows are available in modern ATE software programs. Each window will have its own advantages and disadvantages. Usually, the empirical approach of trying a few different window types on your sampled data can be quite effective. Since most signal processing software packages have the equations and algorithms for implementing these windows, they will not be presented in detail. Instead, a few examples will be given. Figure 6.9(c) shows the original sampled signal multiplied by a Hamming window. The FFT of this windowed sample is shown in the bottom right panel. The noise floor is noticeably lower both below and above bin 5 in the frequency domain.

The next example shows the same signal as before, with a Blackman window applied. In this case, the Blackman window lowers the noise floor even further. Notice in Figure 6.9(c) that like the Hamming window, this one ensures that the amplitude at the beginning and at the end of the sample is very close to zero. Also notice that for this example, the Blackman window has a slightly broader spike around sample 5 in the frequency domain, as shown in Figure 6.9(d). The Blackman window does a good job of lowering the noise floor, but it has more close-in effects on the signal of interest (Figure 6.10).

What about the case where coherent samples are possible? Figure 6.11(a) shows the case where we have exactly five cycles of a sinusoid. The starting phase is identical to the ending phase, and coherence is maintained. The FFT of this waveform has a very low noise floor of less than −80 dB, as shown in Figure 6.11(b). The noise

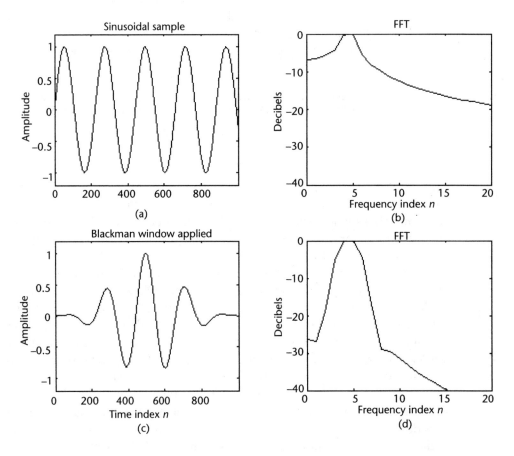

Figure 6.10 Noncoherent sampling and Blackman window: (a) noncoherent time domain, (b) noncoherent frequency domain, (c) windowed time domain, and (d) windowed frequency domain.

6.6 Dynamic Measurements   159

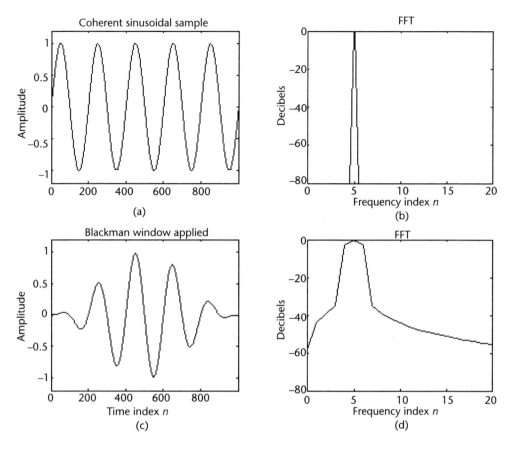

Figure 6.11 Coherent sampling and Blackman window: (a) coherent time domain, (b) coherent frequency domain, (c) windowed coherent time domain, and (d) windowed coherent frequency domain.

floor could vary quite a bit depending on the instrumentation, but for this illustration, −80 dB is pretty close to what is seen in many real-world applications. Does using a window help at all in this case? Figure 6.11(b, c) shows the sampled waveform and its FFT results. Applying a Blackman window to the coherent signal and finding its FFT shows that, as we expect, the Blackman window widens the signal of interest in the frequency domain. Unfortunately, the noise floor has been raised noticeably higher than the coherent sampled FFT. This tells us that whenever possible, we should use coherent sampling. In the case where we do not know the frequency of the signal of interest or it has complex modulation, windowing can help us to lower the noise floor, but only to a point.

6.6.2 SNR for AWGs and Digitizers

In the frequency domain, the signal-to-noise (SNR) ratio can be found by comparing the desired signal to the noise floor of the rest of the signal. This is illustrated graphically in Figure 6.12. The SNR does not include the effects of the harmonics of the signal of interest.

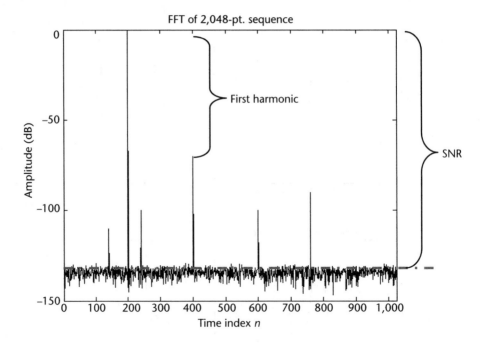

Figure 6.12 SNR and harmonics.

6.6.3 SINAD and Harm Distortion

Another dynamic measurement that is commonly used is SINAD, which stands for signal to noise and distortion. The equation for finding this is

$$SINAD = 20\log_{10}\left(\frac{amp_{desired}}{\sum\sqrt{Other\ bins^2}}\right) \quad (6.9)$$

SINAD does include all other components of the spectrum, the spurs, the harmonics and the noise. SINAD is calculated in the frequency domain, so in the case where our desired signal is in bin M, simply compare the amplitude in bin M to the RMS sum of the values in the other bins. Finding $20\log_{10}$ of this number gives the result in decibels.

Harmonic distortion compares the signal in bin M with the signals in bins kM where k usually equals 2, 3, 4 ... up to 7 bins. This is calculated by (6.10).

$$THD = 20\log\left(\frac{\sqrt{\sum(KM)^2_{k=2,3,4...7}}}{\sqrt{M^2}}\right) \quad (6.10)$$

Let's try and tie together some of these concepts that have been introduced. Starting with the example shown in Figure 6.13. This signal is a simulation of a true time-domain signal that might be seen in a real-world device. The original time-domain sequence contains 2,048 points. The signal is a cosine wave, which contains a total of 200 cycles in the sequence of 2,048 points. Added into this signal for effect

6.6 Dynamic Measurements

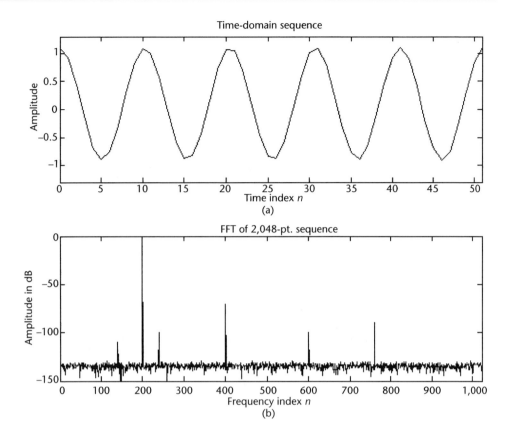

Figure 6.13 Coherently sampled signal.

are two harmonics, a dc offset, noise to show the noise floor of a typical device, and a few spurs. Note that in the time-domain signal, most of these additive signals are not even visible. Finding the FFT of the array and converting it into decibels, these signals become apparent. Note that since we have acquired 200 cycles, the dominant tone is in bin M. Bins 400 and 600 contain the first two harmonics of the signal. Relatively large spurs can be seen at bins 150, 230, and 760.

Figure 6.14 shows the effects of slightly changing the frequency of the sinusoid without compensating the sampling rate; the rule of coherence has now been violated. The top plot shows the frequency spectrum using a rectangular window or no windowing. Losing coherence wreaks havoc on the noise floor. Looking at the plot, the noise floor is so high that all components of the spectrum are lost except for the dominant tone and the first harmonic. The SINAD for this plot is only 6.02 dB. Even though the spur at location 760 is quite high, it is not even visible and is indistinguishable from noise. This is a classic case where windowing could be quite useful.

The bottom plot in Figure 6.14 shows the exact same time-domain sample with a Hanning window used before the FFT. The use of the Hanning window drops the noise floor considerably such that the SINAD has jumped to 23.4 dB. This is extremely useful since now many of the other components of the spectrum become visible. The first and second harmonic can now be seen, as well as the spur at bin 760. There is still quite a bit of smear of the tones as compared to the original

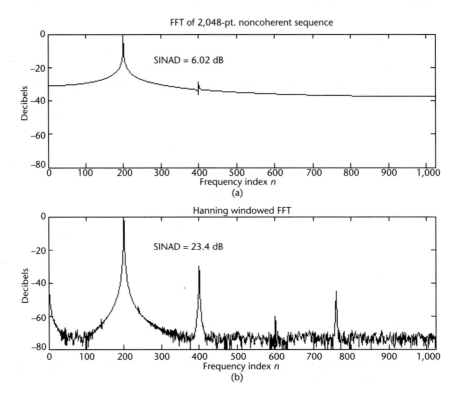

Figure 6.14 Rectangular and Hanning window for noncoherent sample.

coherent sampled case, but here the Hanning window clearly improved the frequency representation of the signal.

The Hamming window was also tried on this sequence. Figure 6.15 shows the results of that experiment. For this window, the noise floor is noticeably better than

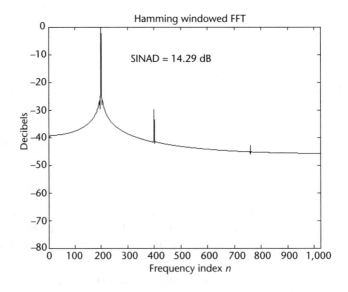

Figure 6.15 Noncoherent sampling with Hamming window.

the no-window case, but the only features that are visible are the first harmonic and the spur at bin 760, and the SINAD was calculated at 14.29 dB.

The next attempt was made using a standard triangle window. A triangle window is exactly what it sounds like in the time domain; it is equal to 1 in the center of the waveform, and 0 on the ends, with a straight line between the center and the end points.

As Figure 6.16 shows, the triangular window is also a notable improvement on the rectangular window with a SINAD of 13.03 dB. Once again though, only the first harmonic and the spur at bin 760 are visible.

By far, of all the windows attempted on this example, the Blackman window had the best effect on lowering the noise floor. As shown in Figure 6.17, its SINAD was improved to 29.7 dB. Also visible are most of the major components of the original sequence, the first and second harmonics, along with the spur at bin 760. Looking carefully, the spur at bin 230 is visible, while the one at bin 150 is not.

6.7 Static Measurements

In order to test the functionality of an analog block there are a few basic conditions that can also be checked. If an ADC or AWG is treated as a functional block, it becomes necessary to check for stuck bits or nodes, as well as for any linearity errors that may occur. These tests do not need any kind of modulation or high-speed analog input. In fact, these measurements are completely separated from timing requirements.

6.7.1 DC Offset

The first and most obvious is the dc offset. To test an ADC, 0V is applied to the input, and the resulting digitized code is analyzed. If its mean value is significantly

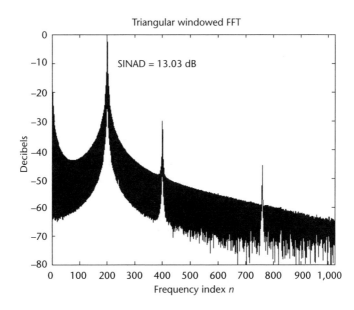

Figure 6.16 Noncoherent sampling with triangular window.

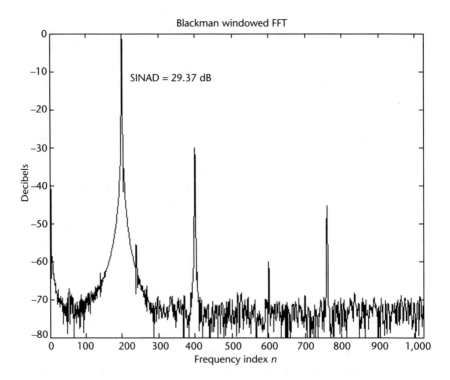

Figure 6.17 Noncoherent sampling with Blackman window.

different from zero, it is said to have a dc offset. For AWGs, the code for 0V is sampled to the DUT, and the resultant voltage is recorded for a period of time. If the output contains a value that is significantly different from 0, then it, too, is said to have a dc offset.

These offsets are usually undesired, but can be compensated for in some instances. If compensation is possible, the voltage or code out is used to calculate an offset that is programmed into an internal register, and the test is rerun.

6.7.2 INL/DNL for AWGs and Digitizers

To check the functionality of all of the AWG or digitizer states, each code must be checked. If the device contains 12 bits, this means a total of 4,095, or 2^{n-1}, conversion codes need to be tested. Luckily, there is an easy way to do this. For the case of a digitizer, a ramp is programmed into the ATE's AWG. The ramp will start from the lowest input value and rise linearly up to the highest input value. This range of voltages is called the full-scale range (FSR). The digitizer is synchronized to perform exactly 2^{n-1} samples starting at the very beginning of the ramp, all the way to the end. Synchronization is very important to ensure the full scale is recorded. In the case of an AWG, a digital pattern is sent to the device that will cause it to step through all of the different codes from lowest to highest, resulting in a ramp typically ranging from $-FSR/2$ to $FSR/2$ on the output pin. The ATE then digitizes this ramp and analyzes the results. Even though this ramp changes in time, this is still called a static measurement because it does not matter how quickly the ramp rises, it just matters that the ATE and the DUT are synchronized.

The first static measurement is called differential nonlinearity (DNL). After the conversions have taken place, detailed calculations are made that compute the difference between the actual step size and the ideal step size. Ideal step size is merely the full-scale range divided by the total number of codes. Once this array is complete, the worst-case or largest DNL is recorded.

The next static measurement called integral nonlinearity (INL) can be calculated from the same resultant array as the DNL. The starting and ending points are recorded, and an ideal straight line is made through these points. For each sample, the deviation from this straight line is recorded as shown in Figure 6.18. Once again, the worst-case deviation is recorded for the device as its INL level. Most modern ATE systems have built-in algorithms for calculating these two static measurements, but suffice it to say that INL looks at deviation from a straight line, and DNL looks at deviation from the step size.

6.8 Real Signals and Their Representations

Looking at specification sheets for RF SOC devices can get pretty confusing. All kinds of terms can be found to describe signals and power levels. It is also common to find mistakes in specification sheets because of fundamental misunderstanding of units. This section will attempt to reduce some of this misunderstanding and provide some useful ways to translate common units from dBm to volts and vice versa.

6.8.1 Differences Between V, W, dB, dBc, dBV, and dBm

In RF systems, the most commonly seen terms are dBm and dB. Let's discuss these terms a bit to make sure that it is clear what they represent. The term dB stands for decibel, and it is merely a scaling factor, which can be added to a ratio of some sort. When the term dB is used, it always compares one signal to another. Often, it compares one power level to another. Whenever the term dB is assigned to a signal,

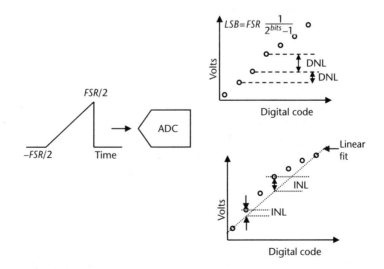

Figure 6.18 Integral and differential nonlinearities.

make sure that you understand what the signal in question is and what signal it is being compared to:

$$dB = 10 \log_{10}\left(\frac{A}{B}\right) \tag{6.11}$$

When signal B is a carrier wave or desired signal, the ratio now represents dBc, or dB referred to the carrier level. If signal B is 1 mW, then we are referring to dBm. If signal A is a voltage (usually rms) and signal B is 1V, we are referring to dBV. However, when we are using units of voltage, there is a scaling factor added such that

$$dBV = 20 \log_{10}\left(\frac{V_{meas}}{1_{volt}}\right) \tag{6.12}$$

This is done because of the relationship between power and voltage. Recall that

$$P = \frac{V^2}{R} \tag{6.13}$$

In the case of comparing powers, we have a V^2 term in the numerator and denominator. They cancel each other out and therefore we multiply the logarithm by 10. In the case of comparing voltages, for consistency, we need to be sure that the squared term is not lost. To do this we multiply the scaling factor by 2 to make 20. This is done for uniformity so that a signal, which is 10 dB below another signal, will always have the same power ratio, regardless of whether we use dBm or dBV.

6.8.2 Transformation Formulas

One common problem for RF SOC engineers is connecting a mixed-signal instrument to an RF port. When setting up an RF source, or making an RF measurement, units of dBm are usually used for uniformity. If we would like to create a signal using an AWG, and all we know is its required power, how do we find out what its peak-to-peak voltage swing should be?

$$V_{RMS} = \sqrt{R \times 1 \; mW \times 10^{\frac{P_{dBm}}{10}}} \tag{6.14}$$

In (6.14), dBm represents the dBm value that we are trying to convert, and R stands for the impedance of the interface, which is usually 50 ohms, but can be different. In order to convert the V_{rms} into peak-to-peak value, multiply V_{rms} by 1.414 for V_{peak}, and then by 2 for $V_{pk\text{-}pk}$.

What about the case when we need to hook up an instrument that has a voltage control, but no power setting? How can we convert from voltage to dBm?

$$dBm = 10 \log\left(\frac{\frac{V_{rms}^2}{R}}{1\,mW}\right) \tag{6.15}$$

For (6.15), the R stands for the impedance at the interface, which is typically 50 ohms, and the voltage is in RMS.

6.9 ENOB and Noise Floor: Similarities and Differences

One of the key sticking points that come between RF and mixed-signal engineers is the way that system noise is described. RF engineers are interested in noise figure and noise floor. Chapter 8 gives a good overview of noise and the noise floor from the RF point of view. Mixed-signal engineers are more interested in the dynamic range of their instruments. In fact, mixed-signal engineers are so interested in the dynamic performance of their AWGs and digitizers that they created a term, *effective number of bits* (ENOB), to describe the real operating range of an instrument. Analog engineers realized that by adding more and more bits to a digitizer, for example, they were able to get more and more dynamic range and, thus, a higher SNR. The problem was that not all 14-bit digitizers were alike. Some had spurs or other noise problems not related to the noise floor of the universe (–174 dBm/Hz.) ENOB shows not so much the number of bits in the instruments, but the number of effective bits.

$$ENOB = \frac{(SNR_{dB} - 1.761)}{6.02} \quad (6.16)$$

Looking at (6.16), we can see that the key here is the signal-to-noise ratio of the signal in question. So, a digitizer that has 14-bits may have only about 12.4 effective bits. Unfortunately, this is still kind of vague. There are all sorts of games that can be played to make an instrument seem better than it really is. Pay attention to operating bandwidth and input voltage swing when looking at ENOB specifications.

Tying ENOB to noise floor can be pretty straightforward after looking at the same equation from a different point of view:

$$SNR_{dB} = 6.02 \times ENOB + 1.761 \quad (6.17)$$

If we know that a digitizer hooked to a 50-Ù load has a total of 10.4 effective bits for a bandwidth of 10 kHz, this tells us that it has a signal-to-noise ratio of 64.36 dB. For the sake of illustration, let's assume that the input voltage signal is 1V peak to peak. In a 50-Ù load, that becomes 3.9 dBm. So, the peak signal is 3.9 dBm, and the SNR is 64.4 dB, which puts the noise floor at –60.5 dBm over 10 kHz. To find the noise floor in dBm, just subtract $10 \log(BW)$ or 40 dB. Now we have found our noise floor from ENOB to be –100.5 dBm/Hz.

Knowing this relationship, we can also work back the other way. If the noise floor is known, the bandwidth of interest is defined, and the peak amplitude of the signal in question is also known, the ENOB can be calculated directly using (6.16).

6.10 Phase Noise and Jitter

Two other sticking points between mixed-signal and RF engineers are the concepts of phase noise and jitter. These two concepts are deeply related, but different in many ways. Both of them can describe the reliability of a clock signal, but they do it in different ways.

6.10.1 Phase Noise and How It Relates to RF Systems

RF systems are built by combining signals in the frequency domain. Upconverting and downconverting are commonplace. Whenever up- or downconversion occurs, any noise in the reference LO will bleed into the resultant spectrum. RF engineers describe this noise as phase noise and measure it in dBc/Hz at specific offsets from the carrier. To measure this value, a source is turned on, and a plot is made in the frequency domain. The reference value of the source is recorded, and then at a specific offset, the noise is recorded and converted into dBc/Hz. The most important point about computing phase noise is that a reference power, an offset, and a bandwidth in which the noise is measured are needed. All of these components are needed in order to compute phase noise.

6.10.2 Jitter and How It Affects Sampling

Mixed-signal engineers are extremely interested in the accuracy of their samples in the time domain. Any error in time of when exactly a sample occurs is called jitter. Typically, in mixed-signal systems, the digitizers are clocked by a digital waveform. Recall from Fourier analysis that a clock has a pretty wide frequency response, and therefore, phase noise is not possible to measure. What can be done is that a jitter analyzer can be used to look at a large sample of edges of a digital waveform, and it can calculate statistically what kind of variation is occurring in the edge placement. This is typically measured in picoseconds for today's state-of-the-art devices. Any jitter on a digitizer or AWG's clock will introduce noise into the measurement, raising the noise floor and reducing the SNR.

The thing that ties these two concepts together is that ultimately clock references are analog signals. Any analog signal with a very good phase noise will be downconverted into a clock with very good jitter performance. Thus, any system with low jitter will ultimately be sourced by a low phase-noise source.

6.11 I/Q Modulation and Complex FFTs

From Fourier analysis we learned that real time-domain signals that are converted to the frequency domain have an even frequency response. This means that in the frequency domain, the positive and negative frequency plots are mirror images of each other. In the case of a time-domain signal, which has real and imaginary components, this is not the case. The positive and negative frequency plots might be totally different. In fact, modern RF modulation systems depend on this fact. This is why most digital radio systems with complex modulations represent baseband signals with both I and Q components. I and Q stand for in-phase and quadrature. In the Fourier domain, this means real and imaginary. The I and Q signals are orthogonal to each other, which means that mathematically we can treat them as complex time-domain signals. If we sample both I and Q at the same time, we can combine both the I and Q arrays to create one complex array that can be processed and analyzed for all kinds of properties.

6.11.1 System Considerations for Accurate I/Q Characterization

Keep in mind that if we are trying to sample both I and Q signals from a radio or to source I and Q signals into a radio, in order for the math to work, orthogonality must be maintained. This means that the amplitude response of our instruments must match each other exactly. Also, samples must be made at precisely the same time for both I and Q signals. Any variation in our instrumentation amplitude and phase will definitely affect the DUT response. Consider characterizing the ATE system thoroughly before beginning work in the complex domain.

In measuring DUT performance for I and Q, we become particularly interested in phase and amplitude match of the DUT. How closely do the I and Q paths in the DUT reproduce the same signal? One method to test receivers would be to apply a single tone to the input. This will be converted to baseband as a single tone at both I and Q. These two tones can then be sampled and compared for amplitude and phase. In I/Q digital receivers, they should have the same amplitude, but have a 90° phase shift between them.

6.11.2 Amplitude and Phase Balance Using Complex FFTs

There is another way to perform this kind of amplitude and phase mismatch analysis. Using a complex FFT, both phase and amplitude balance information are embedded in the measurement itself, if you know what to look for. Figure 6.19 shows the first 11 samples of I and Q signals coherently sampled, such that 200 cycles have been captured. A total of 2,048 points were taken for both I and Q. The plot with the diamond symbols is the I-plot or real component of the array. The other one, with circle symbols, is the Q, or imaginary, component from the simulated receiver. Note that the imaginary, or Q, signal leads the real signal by 90° and the peak input voltage is 1.0V.

The bottom of Figure 6.16 shows the complex FFT. In all previous discussions, if we sampled N points in the time domain, we only looked at $N/2$ points in the

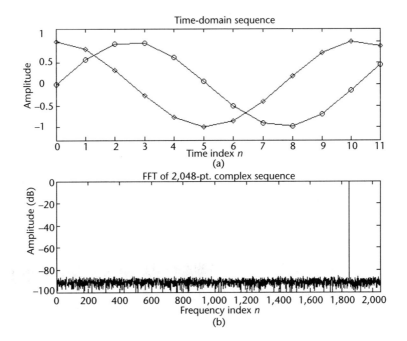

Figure 6.19 Complex time domain and complex FFT.

frequency domain because we knew that the positive and negative frequencies were identical. This is only true for real time-domain signals. In the case of complex time-domain data, we are interested in all N points because we know that the positive and negative spectra are different, and now we want to compare them to each other. Note that the frequency response shows a single peak at bin 1,848 (2,048 − 200). This is equivalent to having a peak at the negative frequency, but not at the positive frequency. Working through the math behind the scenes, the peak shows up on the negative plot because the imaginary signal leads the real signal in the time domain.

Now that we have found the FFT and found the peak, what else can we determine from this plot? It turns out that there is information in the signal at bin 1,848, but there is also information in the fact that little or no signal is present in bin 200.

What happens if a bit of amplitude imbalance is introduced into the signal? A simulation was run with I and Q imbalanced by 0.0001%. For a 1-V peak signal, this would mean less than a 1 μV of amplitude imbalance. Notice that in Figure 6.20 information is now seen in bin 200. Since the signal in bin 200 is more than 50 dB down (i.e., 50 dBc) from our desired signal, we know that the contribution due to amplitude imbalance is low; however, this illustrates the point that there is no need to perform a separate amplitude imbalance calculation. All that is needed is to measure the decibel difference between bins 1,848 and 200. Since the difference is greater than 50 dB, in this case we know that the amplitude imbalance is less than 1 μV.

Next, the simulation performs an absolute match in amplitude while introducing a very slight phase imbalance. Figure 6.21 shows what happens to the original complex signal with an amplitude imbalance of 0.001°. Note that once again, phase error is visible in the signal in bin 200. Remember that an ideally matched complex FFT of orthogonal sinusoids will not have a component in this bin. For this example,

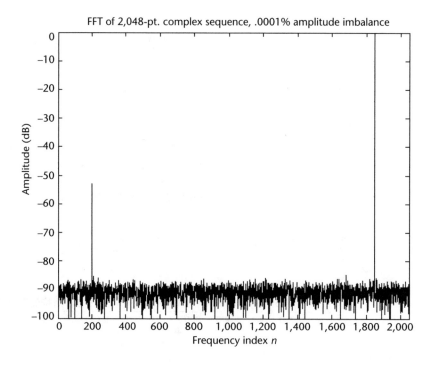

Figure 6.20 Complex FFT with amp imbalance.

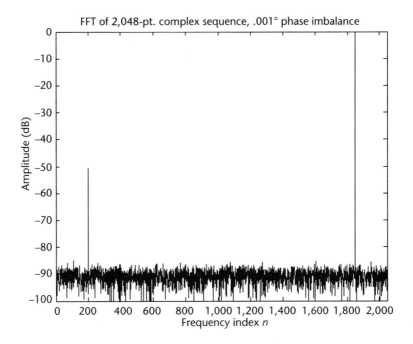

Figure 6.21 Complex FFT with phase imbalance.

this very slight phase error appears as a signal that is 50 dB down from our desired signal.

This example shows that by taking the complex FFT, both the amplitude and phase match can be measured in decibels by looking at the difference between the positive and negative frequency component bins of our FFT. Our test should simply ensure that the amplitude difference between bin 200 and bin 1,848 is greater than 50 dB to ensure both phase and amplitude match are maintained.

6.12 ZIF Receivers and DC Offsets

6.12.1 System Gain with Dissimilar Input and Output Impedances

As RF SOC devices increase their integration, and designers mix and match the kinds of blocks used, test engineers are no longer guaranteed uniformity in the system impedances. In fact, it is quite common for an RF system to have different impedances at each block in order to optimize battery life. Some output blocks are even meant to go into a high impedance buffer with greater than 1-MΩ load impedance. How should the test engineer handle the case of a system with a 50-Ω input impedance in RF, but a different or high impedance at baseband? Consider the block diagram in Figure 6.22.

In this case, if we know the level of the signal that goes into the device, how should we go about measuring system gain? One approach might be to convert the output level into power and then dBm and subtract. The problem with this approach is that the baseband signal is being measured with a digitizer in the ATE. Any signal incident on a 1-MΩ load will have a very small variation in V for relatively large variations in power. Solving the power equation for V gives

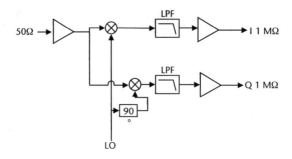

Figure 6.22 Ziff receiver with different in and out impedances.

$$V = \sqrt{\frac{P}{R}} \qquad (6.18)$$

If R is very large, V will be relatively unaffected by variations in P. It is difficult from a practical viewpoint to measure gain this way.

Perhaps a better approach would be to convert the input power back into voltage using (6.14.) Next, measure gain directly by comparing input V_{rms} to output V_{rms}. Find 20 multiplied by the log of this ratio. Gain has now been computed quickly and accurately.

6.13 Summary

The language of mixed-signal engineers is very similar to that of RF engineers. Sampling a continuous waveform and analog reconstruction of that signal is of primary concern to mixed-signal engineers. Understanding the link between the time domain and the frequency domain is key. The mathematics of mixed-signal engineering is the Fourier series, the Fourier transform, the discrete Fourier transform, and ultimately the FFT. Several examples were presented that showed the relationships between the continuous time domain and the continuous frequency domain. The DFT is merely an approximation of that ideal, and the conditions required for this discrete transform were presented. Coherent sampling is the best way to capture a signal, but using a variety of windowing techniques can greatly improve a signal's SNR where coherence is not possible.

Analog and RF engineers use many of the same terms to describe signals such as SNR and SINAD. This chapter showed these same concepts from a mixed-signal or sampled voltage point of view. Many of the terms like decibels, watts, and volts are familiar to both RF and analog engineers, but dBV, dBm, and dBc were more closely scrutinized to make sure that conversions from one unit to another, where appropriate, could be understood. Noise floor is ultimately the limiting factor for dynamic range in RF engineering, while ENOB is the benchmark for mixed-signal engineering; these concepts along with phase noise and jitter were also explained. After the basics were introduced, a few cross-domain examples were given to tie both realms together. The example of the complex FFT showed how I/Q modulators can be characterized using basic mixed-signal techniques. The gain measurement example showed how gain can be calculated using appropriate unit translation. These basic concepts and cross-discipline examples should help to further the reader's understanding of how to handle complex SOC designs that convert signals from RF to baseband and baseband to RF.

References

[1] Ramirez, Robert W., *The FFT Fundamentals and Concepts*, Upper Saddle River, NJ: Prentice Hall, 1985.

[2] Bracewell, Ronald N., *The Fourier Transform and Its Applications*, 2nd ed., New York: McGraw Hill, 1986.

[3] Soft Test, Inc., "The Fundamentals of Mixed Signal Testing," 2000, at www.soft-test.com.

CHAPTER 7
Moving Beyond Production Testing

7.1 Introduction

Cost of test (COT) is an ever-increasing and important issue among chip vendors. The longer it takes to test a chip, the lower its throughput and the higher its production cost will be. Much of what motivates chip manufacturers and test houses to employ parallel and concurrent testing methods is the attempt to reduce the overall COT (see Chapter 5 for more detail about COTS). One good way of reducing the overall test time or the effective test time is to test chips in parallel. This has a direct positive impact on COT because in theory it increases the throughput as compared with the test time of single-site testing. Most chip vendors employ some sort of parallel testing, especially if the production chip volumes are high or the market life of the production chip is or will be very long [1].

7.2 Parallel Testing of Digital and Mixed-Signal Devices

Parallel testing of digital and mixed-signal devices has long been the norm. Digital device testing is naturally conducive to parallel methods, and 16-, 32-, and 64-bit parallel buses have been in existence for decades. If one device is already being tested with a 64-bit parallel bus, then why not test a second, or third, or fourth device in parallel using the same method. Figure 7.1 shows a block diagram of two devices being tested in parallel. Using this simple digital test model, the number of devices that can be tested in parallel is limited only by the number of digital pins available on the test system.

Historically, the speeds of such devices were in the megahertz range. These slower speeds allowed parallel test interfaces, load boards, and test boards to be built with relative ease. Mixed-signal devices have multiple analog-to-digital converters (ADCs) and digital-to-analog converters (DACs) on a single chip, so it is a natural progression to consider testing such devices using multiple digitizers and multiple arbitrary waveform generators, and indeed, this is the norm of today's test industry [2].

7.3 Parallel Testing of RF Devices

Parallel testing of RF devices is not nearly as common, though there is increasing pressure to follow the parallel digital path. There are several reasons for the test

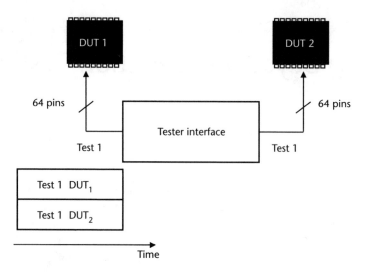

Figure 7.1 Two devices being tested in parallel.

industry's infancy regarding RF parallel testing. There is a fundamental difference between the RF devices and their digital/mixed-signal counterparts. The fundamental difference is the radio. Its unique function prevents standardization and hinders blind parallelism. In the RF and wireless arena, everything coexists simultaneously because it is in the open air. Right now as you read this page, signals across the entire spectrum (including, but not limited to RF) are passing through and around you. The function of the radio is entirely unique, as it must select only the desired signal while rejecting all other signals. It is the radio's uniqueness that insures that your mobile phone does not pick up TV stations or broadcast over the local radio station. Additionally, the radio must meet the link requirements imposed by the atmosphere, antenna designs, and transmitter/receiver spacing. Once the desired signal is downconverted and processed into digital bits, any bit is as good as the next. A TV bit, FM radio bit, or mobile phone bit can all be identical, but 250 mW at 5.6 GHz will not meet the link margin of a radio that needs 1W at 900 MHz. Radios must be unique to avoid interference, and their uniqueness hinders standardization. Until recently, RF devices were mainly built using a single-radio architecture. The radio architecture housed one or more of the basic RF building blocks such as amplifiers, mixers, VCOs, PLLs, and filters. There was no digital [other than a digital control bus, usually in the form of three-wire serial protocol interface (SPI)], and there were also no mixed-signal components such as ADCs or DACs. In addition, the high frequencies (in the gigahertz region) and broad dynamic range requirements (greater than 100 dB) of RF devices make designing and building parallel RF test interfaces much more difficult.

More importantly, RF testing and RF testers have remained a systems-integration endeavor, whereas digital and mixed-signal testers are further along in the card-based integration arena. Digital cards for digital testers are built on wafers in very much the same way that digital devices are. In contrast, RF test systems are unique, like the radio devices that they intend to test. The various digital cards with a wide range of functionality can easily be designed in and mixed and matched to a particular test system. Mixed-signal digitizers and arbitrary waveform generators are being built with multiple cores and multiple inputs and outputs. These multiple cored mixed-signal testing units can be viewed as mixed-signal semicards. That is, the design and form factor of these semicards is approaching the digital-like card level, with the next level being a mixed-signal card that is very similar to a digital card, but it is not quite there yet. Currently and unfortunately for the test industry, RF testing is at an integration level that is still largely at or near a systems level. That is, a large percentage of RF systems are built by integrating RF building blocks. For example, RF sources are manufactured by a variety of vendors that all specialize in the design and manufacture of RF sources. These sources are purchased and integrated into an RF test system. Not much of RF technology is card based or even semicard based. Attempts are being made to improve upon this, and certain pieces of RF test systems can be thought of as modular or semicard-like (for example, an RF switch matrix or IF MUX), but there is still a long way to go.

The integration of multiband radios onto a single chip is pressuring the development of parallel RF testing methods similar to digital to take advantage of this multiple-radio architecture. For example, an SOC WLAN card might have three radios integrated into it to cover 802.11a, 802.11b, and 802.11g. The other nonradio functions of the SOC can be tested in parallel, which creates a test-time bottleneck for the RF testing. Another example might be a cell phone having multiple radios to cover various geographical areas around the globe, as well as a Global Positioning Satellite (GPS) radio and even a Bluetooth radio.

It is important to note that in both SOC examples, there may only be one processor or one memory chip or one microcontroller, but there are multiple radios that are required. RF ICs are fundamentally different from digital ICs. This fact is often overlooked by the industry and can lead to poor testing decisions. Although there are trends and strategies from the digital world that RF fabs can learn and benefit from, the uniqueness of radios is the fundamental reason RF integration levels do not track digital integration levels. Higher levels of integration are a trend that has worked to the advantage of both the system application and semiconductor fabs for digital circuitry. It has not worked as well (and may never work) for RF circuitry for many reasons that are fundamental and unique to the RF system design problem. It has been especially disruptive and costly to semiconductor fabrication facilities that specialize in RF circuitry. Certainly, higher levels of integration will continue to be sought by manufacturers of RF systems, and over the course of a particular product's history, there will be a trend toward integration. But when the next new product is released, all components (LNAs, filters, PAs, mixers, duplexers, and so forth) require a complete redesign. The power, frequency, linearity, modulation scheme, and matching circuitry must all be designed from scratch. The risk and design interleaving time to realize the new circuitry in a completely integrated fashion drives the new product toward a more discrete realization—initially. Another factor that pushes the design in this direction is the fact that the optimum technology to achieve RF performance for each RF function is different. High levels of integration sacrifice performance by forcing all functions into a single technology. Testing is often made more complex and expensive by higher levels of integration. This is very different from the digital circuit situation where a bit is a bit, a logic function is a logic function, and so forth. Increasing the level of integration of a digital chip is often simply a matter of taking two complementary metal oxide semiconductor (CMOS) layouts and attaching them at the appropriate points. Testing is actually sometimes made simpler.

As an example, note that the trend toward the integration of RF circuit is not as rapid as the trend toward digital circuits. Going from a clock rate of 900 MHz to 1.1 GHz is a matter of scaling circuits by defined rules. The gain to system performance is immediate. Compatibility is not an issue. Going from 900-MHz analog modulation to 2.1-GHz CDMA changes the whole system, every component and every matching element. It is not possible to scale the design, but rather the new system must be designed from scratch.

7.4 Parallel Testing of RF SOC Devices

The last few years have seen the industry successfully integrate the I/Q modulator to create a complete RF receiver/transmitter. Now that companies have working base designs that have a complete radio modem on a single chip, they are integrating further to build multiple-band radios on a single chip, and indeed, many chip designers are already working on their next generation devices with even higher degrees of integration.

When considering a mobile device such as a cell phone, it makes sense to have multiple band radios. Europe is based on a GSM system, while the United States and Asia have various systems, including GSM, CDMA, TDMA, and others. A consumer would much rather purchase a cell phone that would work all over the world instead of only in a specific geographical location. This problem is particularly acute in the United States with its myriad standards scattered across the continent. Multiple-band radios eliminate this problem and offer the consumer more attractive possibilities. The same is true for the WLAN and Bluetooth standards. Each geographical region allocates its frequency bands independently. It is desirable to have a multiband WLAN card that adheres globally to the various frequency bands. Additionally, some of the WLAN standards are not standards yet, but are still evolving. A WLAN card that adheres to multiple standards allows the chip manufacturer to spread his risk and still be in a good market position should one particular standard dominate the market at a later date. The manufacturing cost of the digital/mixed-signal portion of the chip will likely decrease, but increases in RF functionality increase the cost of the RF portion of the chip and can easily counter the digital savings. This should be apparent from the fact that the RF circuitry physically is made up of multiple radios providing the necessary functions.

With companies scrambling to offer products that adhere to the multitude of frequency bands around the globe, it is easy to see why multiband radios are becoming so prevalent. As the trend continues, more and more digital blocks and mixed-signal blocks will be integrated with these multiband radios. Since most of the digital and mixed-signal blocks are already tested in a parallel fashion, this will further pressure the test industry to offer testing of the RF blocks in a parallel fashion as well.

Loosely speaking, the RF production testing requires approximately the same amount of test time as the mixed-signal testing in a single-site setup. If you recall, we mentioned in Chapter 4 that mixed-signal testing and RF testing have similar signal processing requirements. However, in a quad-site setup, the RF testing requires approximately 400% more time than the mixed-signal testing. This is because the mixed-signal testing is performed in parallel using semicard-like mixed-signal testing equipment, while the RF testing is performed serially with a system that has RF components integrated into it at a systems level. A pictorial representation of a device with both mixed-signal and RF testing requirements is shown in Figure 7.2.

In Figure 7.2, where Test 1 is an RF test requiring an RF receiver and Test 2 is a mixed-signal test requiring some combination of digitizer and arbitrary waveform generator cards. Test 2 time, the mixed-signal test time, remains relatively flat with additional sites, while Test 1 time, the RF test time, increases linearly with each additional site. Test time translates directly to chip cost, and thus, the industry is extremely interested in applying the same digital and mixed-signal parallel techniques to RF testing.

7.5 True Parallel RF Testing

Figure 7.3 is a dual-site block diagram of a wireless SOC device with multiple RF inputs and outputs, as well as analog or digital inputs and outputs.

The device will have both an RF input and RF output and could possibly have multiple RF inputs and outputs, depending on the integration level (i.e., if it has multiband capability or not). Depending on the design, the RF input/output could be on the same pin and switched internally by the device, or they could be on separate pins. In either case, the challenge of testing the RF in parallel remains unchanged. The device will typically also have either an analog input/output or digital input/output. Whether it is analog or digital depends on the integration level of the particular design. We have already mentioned that parallel testing methods for

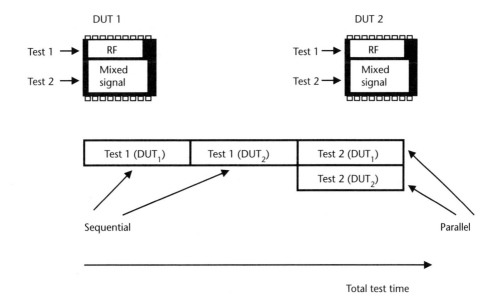

Figure 7.2 Four devices with mixed-signal and RF testing requirements being tested in parallel.

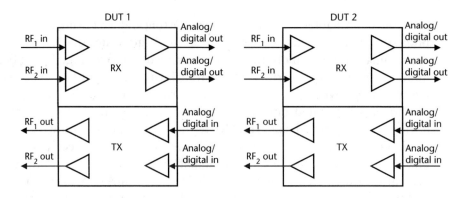

Figure 7.3 Dual-site block diagram of a wireless SOC device.

both digital and mixed-signal signals are quite prevalent; therefore, we will limit our concern and focus to the RF input/output.

True parallel testing is the most efficient method of parallel testing. It also requires the most capital investment from both the test equipment manufacturer and test equipment purchaser because both must purchase duplicate sets of hardware. True parallel testing means that exactly the same tests with the same test conditions are performed simultaneously on all sites. This form of testing requires that duplicate sets of resources are available to perform the tests in parallel. Since RF testing requires some sort of downconverting architecture, two complete downconverting architectures must be available to perform simultaneous dual-site RF testing. Figure 7.4 shows what a true parallel testing architecture might look like for a dual-site implementation.

7.6 Pseudoparallel RF Testing

Most test equipment manufacturers, chip manufacturers, and test houses are currently employing a combination of parallel RF testing and serial RF testing. This combination of parallel and serial RF testing is a kind of pseudoparallel RF testing.

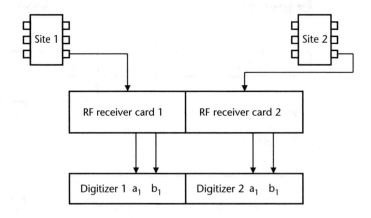

Figure 7.4 Possible true parallel testing architecture.

7.6 Pseudoparallel RF Testing

Pseudoparallel RF testing requires only a single downconverting architecture. The trade-off is that the required capital investment is much less, but the throughput is not as high as a true parallel testing setup. There are many forms of pseudoparallel RF testing. Figures 7.5 and 7.6 show two examples.

In Figure 7.5, the same RF signal is simultaneously stimulating both inputs of Site 1's and Site 2's receiver input chain, and the digital outputs of both DUTs are measured in parallel. (Note: That the outputs are digital is immaterial. They could just as easily be analog outputs with no loss of generalization.) In contrast, since

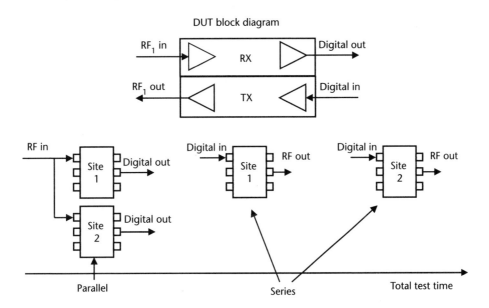

Figure 7.5 Pseudoparallel testing setup.

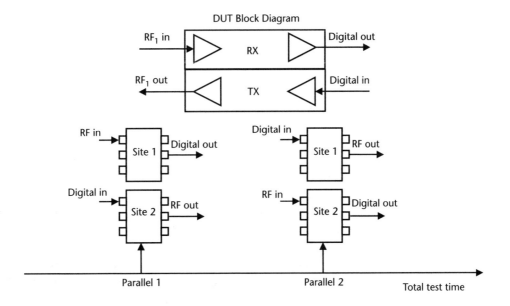

Figure 7.6 Concurrent-pseudoparallel testing setup.

there is only one RF receiver in the test setup, the transmit chains of Site 1 and Site 2 are measured sequentially (i.e., in serial).

Figure 7.6 is a variation of the same theme. RF is still being measured serially across both DUTs. Site 1 is being stimulated with an RF signal while Site 2 is being stimulated with an analog/digital signal. The analog/digital output of Site 1 is measured in parallel with the RF output of Site 2. The stimulus and measurement setup is reversed, then applied again to both DUTs in parallel. Depending on the percentage breakdown of digital, analog, and RF test times, this testing method may yield a better throughput efficiency than the method shown in Figure 7.5.

Performing output RF tests in parallel is generally complicated, and if you do not have a test setup similar to Figure 7.4, it is impossible. Figure 7.5 shows the RF input tests being performed in parallel while the RF output tests are performed serially. Performing input RF tests of SOC devices in parallel is not that complicated because what are actually measured are the output signals. The output signals of a wireless SOC device are either digital or analog. As we already mentioned, parallel digital and analog testing has been in existence for several years now. The RF output tests are generally performed in serial because the test setup has only one RF receiver. The RF receiver is not a card, but rather a system module and cannot easily be duplicated and integrated into a production-test system. One of the main reasons is size. Adding a second receiver module increases the overall footprint of the test system by a large percentage (in the range or 20%), whereas adding extra digital pins or extra mixed-signal cores adds negligibly to the overall footprint size.

7.7 Alternative Parallel RF Testing Methods

Both chip manufacturers and test equipment vendors are constantly seeking new alternatives to reduce the cost of test. Moore's Law, the doubling of transistors every couple of years, has been maintained and still holds true today [3]. More transistors per chip means more functionality per chip. More functionality per chip means more testing is required per chip to test the additional functionality. As the chip size continues to shrink, the test time tends to increase because more and more tests are added for the additional functionality. This has both chip manufacturers and test equipment vendors aggressively searching for new methods to combat this costly trend.

Another method that falls into the parallel RF testing category does not yet have a name and is not being widely employed by chip manufacturers. The method involves applying different RF tests to the individual sites [4]. With this method, the individual site test coverage is equivalent to applying the same tests to all sites, as is currently performed in industry, but the individual sites do not have an equivalent test list. As an example, consider a basic RF power-out test being applied to a Bluetooth radio modem. The Bluetooth standard has 79 different RF channels, but testing all 79 channels would be cost prohibitive, and no chip manufacturer tests all 79 channels in production. Instead, a chip manufacturer usually tests the RF power out of only a few of these channels. The channels are statistically selected to offer the broadest test coverage. That being the case, it is possible and statistically valid to test the RF channels of each site independently and in parallel. In fact, in doing so, the chip manufacturer would obtain a wider set of statistical information about his

7.7 Alternative Parallel RF Testing Methods

manufacturing process. Figure 7.7 shows an example of a quad-site Bluetooth setup employing this alternative method.

Each site's output is routed through an RF combiner, and the combiner's single output goes to the RF receiver. Normally, one would program the radio to a particular frequency and conduct a series of continuous wave (CW) and modulated output tests on the TX chain. Instead, each site can now be programmed to a different frequency, and the test list can be performed in parallel. Thus, the test time will remain relatively unchanged with additional sites tested. This assumes the test equipment has the capability to accommodate the wider bandwidth (BW) requirements. This methodology is very similar to what is done for "go, no-go" testing of digital circuitry, except that with this RF method, most of the data is available for postprocessing if it is desired. Figure 7.8 is a plot of the coupler's output frequency spectrum.

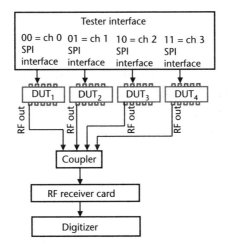

Figure 7.7 Quad-site Bluetooth test setup.

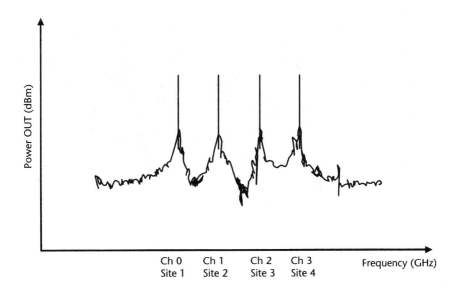

Figure 7.8 CW output spectrum of quad-site Bluetooth setup.

Each channel (1–4) corresponds to its named site number (1–4). Channel 1 is the CW output power of site 1, channel 2 is the CW output power of site 2, and so forth. This spectrum is easily captured and processed by a typical RF receiver. This method is not limited to simple CW power measurements. Figure 7.9 is another example where the measurement is a modulated power measurement.

Again, each channel corresponds to its particular site number. The upside is that this method does not require any extreme capital investment for a second receiver. The major trade-offs are that each site has a different test list, the throughput is not as high as a true parallel setup, and the test system must be able to cope with a more complicated test program and properly log the test data and bin the devices correctly. Application/test/product engineers who are well trained and knowledgeable about wireless SOC testing can design test setups that implement this method successfully.

7.8 Guidelines for Choosing an RF Testing Method

The decision basis for a particular test approach is often dominated by legacy and immediate budget, rather than sound engineering judgment. And it is usually impossible to choose the optimum strategy in the very beginning unless accurate volume predictions of the product over the coming months and years can be made. But it is helpful to understand how to compare the different approaches if the product volume and schedule issues are well defined.

Table 7.1 attempts to provide a guide to choosing a particular RF testing method. The table uses the volume demand and your desired investment method to choose a particular method.

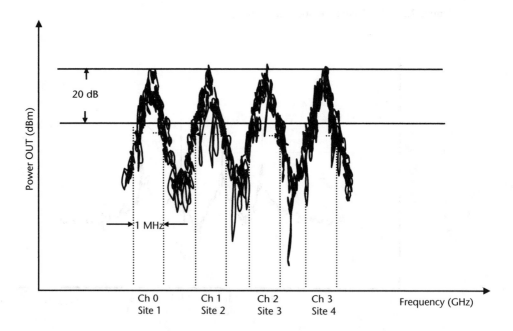

Figure 7.9 Modulated output spectrum of quad-site Bluetooth setup.

Table 7.1 Volume Versus Investment

Volume	Low Investment	Medium Investment	High Investment
Low	Serial	Alternative parallel	Pseudoparallel
Medium	Alternative parallel	Pseudoparallel	True parallel
High	Alternative parallel	Pseudoparallel	True parallel

The investment includes a variety of things, such as capital cost of equipment, load boards, DIB boards, engineering skills, handlers, contactors, test program complexity, and so forth. For low volumes and low investment, it is best to stay with a serial RF testing methodology. If the volume remains low, then perhaps an alternative parallel or pseudoparallel approach can be realized with more investment, but the extra capital cost of a true parallel method is not likely to be justified. As the volume increases, it is likely that medium and high investments for pseudo and parallel testing methods can easily be justified.

Table 7.2 attempts to bound the RF test time-savings that each of the RF testing methodologies offers for dual-site and quad-site implementations versus a serial-site implementation.

This table is just a rough estimate and the percentages can be slightly higher or lower than those listed. Alternative parallel methods provide a throughput advantage over serial RF testing, but with an alternative method, the data will often be in a less efficient format than for a serial site implementation. Pseudoparallel testing offers an even better throughput advantage. This method is likely to be the best choice if the SOC device has many mixed-signal tests in conjunction with the RF tests. Finally, true parallel offers the highest throughput advantage with the theoretical maximum efficiency being 100% per site. However, in practice this is not feasible because of the extra software overhead.

7.9 Interleaving Technique

In most instances a test setup will not have a complete duplicate set of resources so that true parallel RF testing will not be possible. However, there are often many techniques that can be implemented to increase the overall efficiency. One of those techniques is called interleaving. Figure 7.10 shows an example of the test savings that can be realized by employing interleaving.

The basic idea is to better utilize the intrinsic resources of the test system's hardware during the test program execution. In Figure 7.10, during measurement 2, the processor is simultaneously retrieving the data from measurement 1. This process is repeated until the last measurement has been completed. The closer together the measurement time and retrieve times are, the higher the efficiency that is realized.

Table 7.2 Dual and Quad Site Test-Time Savings Bounds Normalized to a Single Site

Number of Sites	Alternative Parallel	Pseudoparallel	True Parallel
Dual (2)	> 100% < 150%	> 150% < 175%	< 200%
Quad (4)	> 100% < 200%	> 200% < 300%	< 400%

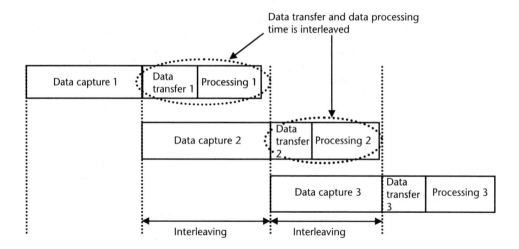

Figure 7.10 Interleaving technique flow chart.

For example, if the data acquisition time is 2 ms and the data transfer and data processing time is 6 ms, then interleaving is providing 2 ms of savings for each measurement. However, as another example, if both the acquisition time and the transfer and calculation times are 4 ms, then interleaving provides 4 ms of savings per measurement. Additionally, an even higher efficiency can be realized if many successive operations of data collection and data transfer and calculation are required, because only the last data transfer is overhead.

7.10 DSP Threading

Another avenue to improve test-time optimization is to employ DSP multithreading [5]. Multithreading is when a computer program can execute multiple branches of code simultaneously. Most programs that are currently being written today are still utilizing a single DSP thread. That is, they execute only a single path in the code at a time. More and more computers and digitizer cards are being built with multiple digital signal processors on board. Digitizer cards that come equipped with multiple DSPs can employ a multithreading technique to further improve test-time efficiency. True parallel RF testing requires multiple RF receivers with multiple digitizers. These digitizers can capture IF signals simultaneously, but their efficiency would be greatly reduced if a single processor then had to process the data sequentially. It would be a waste of hardware and capital investment if each digitizer had to wait on a single processor. That single processor has to acquire the data and perform a plethora of mathematical operations including fast Fourier transforms (FFTs), averaging, resampling, and filtering [6]. These mathematical operations are time intensive and require a greater percentage of the overall time compared to the acquisition time.

Each DSP core or card can be equipped with specific or custom software that will allow a threading technique to be realized; however, this dramatically complicates software development, and even minor changes can adversely impact time to market. Theoretically, multithreading offers the opportunity to realize huge savings on cost of test, but to date the ATE industry has not really attempted to develop this complex technique into production platforms.

7.11 True Parallel RF Testing Cost-of-Test Advantages and Disadvantages

As a reminder, a simple cost-of-test (COT) equation can be written as

$$COT = \frac{(Fixed\ cost + Recurring\ cost)}{(Lifetime \times Yield \times Utilization \times Throughput)} \qquad (7.1)$$

where fixed cost, recurring cost, lifetime, yield, utilization, and throughput are defined as

- Numerator:
    - Fixed cost—Capital equipment cost and floor space cost;
    - Recurring cost—Calibration, cleaning, general maintenance, compatibility cost, contract repair cost, software subscription.
- Denominator:
    - Lifetime—Upgradeability, obsolescence;
    - Yield—Device yield, measurement accuracy, measurement repeatability
    - Utilization—Reliability, uptime, application fit;
    - Throughput—Multisite capability, concurrent testing capability, test time.

Table 7.3 provides a summary of the COT items for the various parallel RF testing methods versus serial.

Obviously, a single site test system requires less equipment; thus, its fixed cost is much lower than that for a true parallel test system. It follows logically that a true parallel test system is more expensive to maintain due to the extra hardware. However, the lifetime of a true parallel test system is greater than a single site system. It stands to reason that if the system is already parallel ready then it should be much easier to add additional sites if required, whereas a single site system may need a

Table 7.3 Cost/Benefit Table of Various Parallel Versus Serial Methods

Cost/Benefit Item	Serial	Alternative Parallel	Pseudoparallel	True Parallel
Fixed cost	Low	Low	Medium	High
Recurring cost	Low	Low	Medium	High
Lifetime	Low	Medium	Medium	High
Yield	Higher	Lower	Lower	Lower
Utilization	High	Medium	Medium	Low
Risk versus volume	Low	Low	Medium	High
Throughput advantage	Low	Low	Medium	High
Implementation complexity	Low	Low	Medium	High

complete overhaul. The yield of any parallel system will be lower than a single site system due to interference issues that are not present on a single site test system. The lower yield impact can be minimized with excellent load board design, but it cannot be completely eliminated. The risk of a true parallel system is that some of the sites may sit idle if chip volumes are not high enough to justify turning the extra sites on. Conversely, if the chip volumes are high, then true parallel offers a considerable throughput advantage over single site systems, with alternative and pseudoparallel systems being somewhere in the middle. Lastly, the implementation complexity increases with additional sites.

Parallel testing improves the cost of test by increasing the throughput. As previously stated, true parallel testing will increase the throughput more than the other parallel method, but it also requires the highest capital investment (i.e., the fixed cost of the test system increases). A disadvantage of true RF parallel testing is that it is a known fact that the yield will decrease on a parallel test cell setup because there is limited real estate on a test interface or load board. This limited space has to accommodate all of the parallel signals at once. Parallel RF signals create many isolation and cross-talk problems, which translate to a more difficult correlation process and ultimately wider test limits. The wider test limits have a direct negative impact on the overall yield. If the chip volume is not high enough, the higher throughput obtained with a true parallel RF setup may not justify the extra capital expenditure and negative yield impact that comes along with it. In many instances, time-to-market is much more critical (in the beginning at least) than throughput (i.e., test time). Chip lifetimes are growing ever shorter (a few months to a year), and it can cost a chip manufacturer or test house millions of dollars to lose a market window to a competitor.

7.12 Pseudoparallel RF Testing Cost-of-Test Advantages and Disadvantages

Pseudoparallel RF testing and the alternative parallel RF testing will have the advantage of increasing the throughput without the additional fixed hardware costs. In addition, with pseudoparallel RF testing there are fewer real estate and isolation issues. Fewer isolation issues means more accurate correlations, narrower test limits, and, ultimately, higher yields. Chip manufacturers invariably have ramping volume requirements. That is, the volume of chips per month demanded of them by their customer increases like a ramp (not necessarily linearly) until it plateaus, begins descending, and then the next chip demand comes along and the process starts all over again.

Figure 7.11 shows the complete life cycle for two devices. Note the two dotted lines on the graph. Volume demands above the first dotted line indicate that more test capacity is required. Further test capacity is required if the volume demands rise above the second dotted line. Note that only for about 50% of the life cycle of device 1 is extra test capacity needed and only for a few short months is "further test capacity" needed. Device 2 has a similar requirement, but with smaller volume demands; only a small portion of the life cycle requires extra test capacity and never does the second device require further test capacity. These

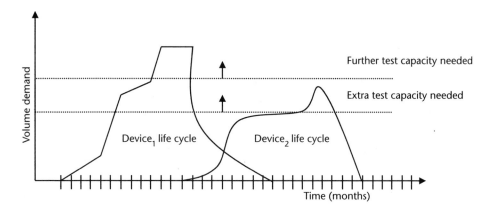

Figure 7.11 Ramping volume graph of two devices.

scenarios are common and make it difficult for the testing company to justify additional test equipment purchases when a majority of the time that "extra" or "further" test capacity will be idle or underutilized.

A ramping volume can mean that the chip manufacturer may need an extremely high throughput for only a few months. In this type of a scenario, it can be much more cost effective to implement a pseudoparallel RF testing technique to cover the peak volume months. This argument is strengthened by the fact that the forecasted (or predicted) volume demands that chip-testing companies receive are about as reliable and volatile as weather forecasts. To mitigate their risk and exposure, while meeting customer demands and expectations, a chip-testing company could employ a pseudoparallel technique to cover the spike months. Thus, the chip-testing company would meet volume demands, have a slightly higher cost of test (but only during the spike months), and would not have to invest in capital equipment that for many months would remain largely idle or underutilized.

7.13 Introduction to Concurrent Testing

Another cost reduction technique that is similar to parallel testing is called concurrent testing [1]. Just as the industry is producing multiband RF chips, it is also producing multiple processors on a single chip. These multiple processors can be tested in parallel if they are designed from the beginning with concurrent testing in mind [7, 8]. Since the processors are imbedded inside a single chip, it is more accurate to refer to this type of testing as concurrent testing rather than parallel testing. Parallel testing is used to refer to the simultaneous testing of multiple chips. Concurrent testing is used to refer to the simultaneous testing of multiple function blocks on the same chip. For example, the simultaneous testing of an ADC, DAC, and Flash memory on the same chip is considered concurrent testing. The simultaneous testing of an ADC on DUT 1 and another ADC on DUT 2 is considered parallel testing.

It is possible to test multiple processors on a single chip concurrently. It is also possible to apply that concurrent test setup methodology across multiple sites in a parallel fashion. In that case the compounded benefits of concurrent testing and parallel testing are realized. A figure of this type of setup is shown in Figure 7.12.

In Figure 7.12, Test 1 is being applied to an ADC. Test 2 is being applied to a DAC. Test 3 is being applied to a CODEC, and Test 4 is being applied to the Flash memory. Test 1, Test 2, Test 3, and Test 4 are all being applied simultaneously to SOC DUT 1. Thus, SOC DUT 1 is being tested concurrently with Test 1, Test 2, Test 3, and Test 4. Additionally, the same Test 1, Test 2, Test 3, and Test 4 are being applied simultaneously to SOC DUT 2. So, both SOC DUT 1 and DUT 2 are being tested concurrently and in parallel with all of the tests. A chip-testing company is able to realize enormous cost-of-test savings (see test-time plots in Figure 7.11) if the test system is capable of both concurrent testing and parallel testing. Moreover, concurrent testing has an advantage over parallel testing in that it allows the test system to continue to evolve with Moore's Law. As new functional blocks are integrated and added to a single chip, the overall test time is limited only by the longest test. Concurrent testing, by definition, then places a lower bound on the cost of test and more accurate cost predictions can be made and better cost-controlling measures can be implemented by chip-testing companies.

7.14 Design for Test

Design for test (DFT) is not a new concept and has been used extensively throughout the maturation of built-in self-test (BIST). However, just as we have seen that parallel and concurrent RF testing methods are behind those of digital and mixed-signal testing methods, RF is also behind in the utilization of DFT and BIST concepts for testing of SOC devices. Progress is being made in both areas, however. For example, Cambridge Silicon Radio (CSR) utilizes BIST technology to test its Bluetooth SOC

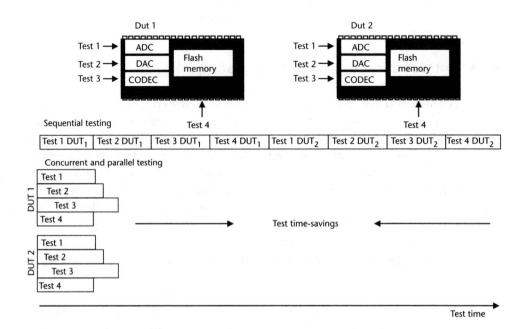

Figure 7.12 Concurrent/parallel testing setup.

devices [9]. Additionally, other companies often employ a receive signal strength indicator (RSSI) in production to test the power levels of their devices [10]. An RSSI circuit is an imbedded circuit on the wireless SOC device that couples off a percentage of the input RF power and converts it to a voltage. The RSSI voltage is a relative indicator of the actual RF power. Thus, instead of measuring the RF power of the SOC directly, one can measure the voltage of the RSSI and indirectly obtain the same results.

Concurrent testing must often be considered before the SOC is actually designed and manufactured. That is, concurrent testing must be included in the design process. Thus, RF DFT concepts are becoming more prevalent with the growing interest in concurrent testing. However, again the dilemma of multiple disciplines occurs. Chip designers in the past have never had to concern themselves with testing, and thus, they often have limited knowledge of test systems and of production-test requirements. Additionally, test/product/application engineers have never needed to understand fully the design side of the chip to produce a production-test solution successfully. Concurrent and parallel testing of wireless SOC devices places enormous, complex demands not only on the test system, but also on the application/test/product engineers who are required to implement such cutting-edge technologies. Not only must these experts come equipped with a wealth of knowledge and be able to understand and apply multiple testing disciplines effectively (digital, mixed-signal, and RF testing techniques) to a single SOC application, but they must perform these responsibilities utilizing parallel and concurrent techniques on highly complex test systems. And they often must work with chip designers where concurrent testing will be realized. Finally, the production solutions must adhere to the time-to-market demands of the chip manufacturer. This is a daunting task at best.

7.15 Summary

The present status of digital, mixed-signal, and RF parallel testing was presented. The integration level of parallel digital testing is ahead of parallel mixed-signal testing, which in turn is ahead of parallel RF testing. Various forms of parallel testing (i.e., true parallel, pseudoparallel, and so forth) were introduced and explained with figures and test scenarios. Other efficiency enhancing techniques like interleaving, and multithreading were introduced. Many of these techniques are used in today's SOC testing solutions, and some of the newer and more complex solutions are just making their way into the mainstream. The concept of concurrent testing was introduced and differentiated from parallel testing. A general COT equation was also provided and used to highlight the trade-offs of the various test setups. Even though parts of RF parallel testing are still in their infancy, production testers with parallel and concurrent testing capabilities are available and being used in production to lower the overall cost of test (or cost of ownership) to chip-testing companies. Constant improvements are being made in both hardware (true parallel RF testing, pseudoparallel RF testing, and so forth) and software (interleaving, multithreading) across all disciplines to lower the capital cost and further reduce the cost of test and cost of ownership. This trend is likely to continue with digital chips to combat the ever-rising cost-of-test issues and to keep up with Moore's Law. As more and more of these digital chips and mixed-signal chips are integrated onto an SOC with multi-band radios, the pressure to implement parallel RF testing methods will increase.

References

[1] Fischer, M., "Concurrent Test—A Breakthrough Approach for Test Cost Reduction," *Semicon Europe 2000*, www.ra.informatik.uni-stuttgart.de/~rainer/Literatur/Online/T/15/3.pdf.

[2] Goto, M., and K.-D. Hilliges, "An ATE Architecture for Systems on a Chip," *Semi Technology Symposium 2000*, Japan, pp. 5-82–5-91.

[3] Moore, G. E. "Cramming More Components Onto Integrated Circuits," *Electronics*, Vol. 38, No. 8, April, 1965.

[4] Schaub, K. "Concurrent-Parallel Testing of Bluetooth/802.11x Chip Sets," *Fifth European Manufacturing Test Conference*, SEMICON Europe 2003, Munich, March 31, 2003.

[5] Shen, J. P., "Multi-Threading for Latency," *Fifth Workshop on Multithreaded Execution, Architecture and Compilation, 34th International Symposium on Microarchitecture*, 2001.

[6] See www.fftw.org.

[7] Mizutani, A., et al., "Concurrent Testing Applied to a Real Device," *SEMI Technology Symposium*, SEMICON Japan 2002, www.ate.agilent.com/japan/stejp/libraryjp/SOC_testjp/CCT_Agilent-Toshiba_E.pdf.

[8] Hilliges K.-D., et al, "Test Resource Partitioning for Concurrent Test," *IEEE International Test Conference 2001*, workshop.

[9] Woolhouse, A., "CSR's New Generation of Bluetooth Silicon Pushes the Bluetooth Market," Cambridge, UK, June 2003, www.csr.com/pr/pr135.htm.

[10] Bardwell, J., "Converting Signal Strength Percentage to dBm Values," at www.wildpackets.com/elements/whitepapers/Converting_Signal_Strength.pdf.

CHAPTER 8
Production Noise Measurements

8.1 Introduction to Noise

Noise is unwanted fluctuation superimposed upon a desired signal. It determines the accuracy and repeatability with which we can measure the signal. During the past few years, improvements in the performance of wireless communications systems have led to the need for tighter specification limits on noise and, thus, a better understanding of it.

Noise is an unfortunate entity that will always be present when performing measurements; for example, an amplifier's output power level is dominated by the noise of the amplifier at very low input power levels [1]. Typically, noise is undesirable, as when noise interferes with a particular parameter that one is attempting to measure, such as a current or voltage signal. In this case noise disrupts the accuracy of the measurement. However, when working with very low-level signals in wireless communications, the need to measure noise levels makes understanding noise desirable. Noise figure and phase noise are two parameters of wireless and SOC devices that warrant an understanding of the behavior of noise.

8.1.1 Power Spectral Density

Noise, being a random process, is characterized as nondeterministic. As a result, when analyzing noise in either the time or frequency domains, statistical approaches must be used. At RF frequencies the analysis is best accomplished using frequency-domain analysis; hence, this discussion will focus on that.

The seemingly obvious approach to characterize noise in the frequency domain is simply to take the Fourier transform of the noise signal. However, this is not possible since the random noise waveform cannot be defined as a simple exact time-domain function. To solve this problem, the power spectral density (PSD) is introduced as

$$S_x(f) = \lim_{T \to \infty} \frac{E\left[|X_T(f)|^2\right]}{2T} \tag{8.1}$$

where E is the expected value and $X_T(f)$ is the Fourier transform of a random noise waveform $x(t)$ evaluated over the time interval $-T < t < T$ [2].

An alternative definition states the power spectral density as being the plot of the power of a signal as a function of frequency as shown in Figure 8.1. The power within a certain frequency range is calculated as

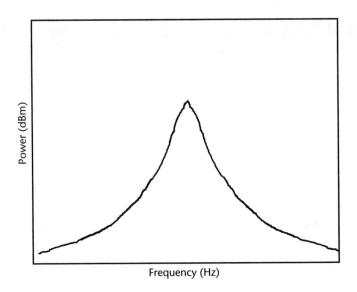

Figure 8.1 Power spectral density for a Gaussian distributed signal.

$$P\big|_{f_1}^{f_2} = \int_{f_1}^{f_2} S_x f(x)\,df \tag{8.2}$$

The total power in a signal is calculated by integrating over all frequencies

$$P_{total} = \int_{-\infty}^{\infty} S_x f(x)\,df = \overline{x^2(t)} \tag{8.3}$$

where $\overline{x^2(t)}$ is a voltage or current signal, and P_{total} is stated as the power across a 1-ohm resistor. As a result of this definition, the units of PSD are V^2/Hz (or more commonly, dBm/Hz when specifically discussing RF frequencies) making stating the bandwidth a necessity when stating the power of a noise waveform. PSD will be used in the following sections to describe the characteristics of different types of noise. The concepts of PSD are also used when discussing noise-figure and phase-noise measurements.

8.1.2 Types of Noise

Noise can arise in many ways. However, within the context of making electronic measurements, noise can be grouped into two types, fundamental and nonfundamental. Fundamental noise consists of that known as white noise, thermal noise, shot noise, quantization noise, and $1/f$ noise. Additionally, in test and measurement systems, nonfundamental noise can arise from electromagnetic coupling, cooling-induced current flow in semiconductors, ground loops due to differing potential reference points, or oscillations in amplifiers. The principle difference between those noise types categorized as fundamental or nonfundamental is the ability to reduce or eliminate the noise (nonfundamental noise) or not (fundamental noise). The figure

8.1 Introduction to Noise

of merit, noise figure, when measured at RF frequencies encompasses principally shot and thermal noise.

Noise exists in many forms, but within the context of testing, the following are the dominant types:

- Thermal noise;
- Shot noise;
- 1/f noise;
- Quantization noise;
- Quantum noise;
- Plasma noise.

Each of these will be discussed briefly to provide detail on their relevance.

8.1.2.1 Thermal Noise

Thermal noise (Johnson noise) is broadband noise resulting from the random motion of electrons due to temperature. The kinetic energy of this random motion is proportional to temperature. This random motion of electrons (charge) produces a voltage across a resistance. It is usually the dominant fundamental noise found in circuits at room temperature. It was discovered by Johnson [3], and the mathematical description was derived by Nyquist [4].

As Figure 8.2(a) shows, thermal noise is of the general white noise class described by equal power spectral density (per hertz) and flat energy across the entire frequency spectrum. White noise gets its name from its analogy with white light, which also has equal power density across all frequencies in the optical band. True white noise cannot exist, as it would require infinite bandwidth by definition, which would also imply infinite energy. A practical description of white noise considers the noise to have a flat power density over some finite bandwidth.

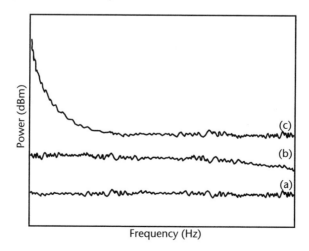

Figure 8.2 Power spectral density.

As the power density of white noise is flat, it is also independent of frequency. This means that white noise signal power for a given bandwidth does not vary, no matter what frequency is chosen, across the entire frequency spectrum. Resulting from this is that white noise in one part of the spectrum is uncorrelated to white noise in another part of the spectrum.

A few fundamental equations describing thermal noise must be introduced at this point. These are the foundation for noise-figure measurements to be discussed later. In 1928 Nyquist derived a formula to describe thermal noise:

$$\overline{v}^2 = \frac{4hfBR}{\left[e^{\frac{hf}{kT}} - 1\right]} \tag{8.4}$$

where $\overline{v}^2$ is the mean-square open-circuit thermal noise voltage across a resistor, h is Planck's constant (6.626×10^{34} J-sec), f is frequency (in hertz), k is Boltzmann's constant (1.38×10^{-23} J/K), T is absolute temperature (in kelvins), R is resistance (in ohms), and B is the bandwidth (in hertz) over which the noise is measured. The derivation of this involves extensive statistical thermodynamics and is beyond the scope of this book. Equation (8.4) is valid for any frequency; however, it is often tedious to work with.

Considering that at microwave frequencies, $hf \ll kT$, the first two terms of a Taylor series expansion can be substituted into (8.4) as

$$e^{\frac{hf}{kT}} - 1 \sim \frac{hf}{kT} \tag{8.5}$$

Substituting (8.5) into (8.4) leads to

$$\overline{v}^2 = 4kTBR \tag{8.6}$$

which is no longer valid over the entire frequency spectrum [due to the approximation of (8.5)]. However, for most microwave/RF work, the approximation and (8.6) are valid and a lot easier to work with. As a worst-case example, consider the case where $f = 100$ GHz and $T = 100$K. In this case, hf (6.5×10^{-23}) is still 100 times less than kT (1.4×10^{-21}). Almost all RF noise calculations are based on (8.6), which is called the Rayleigh-Jeans approximation and is valid unless very high frequencies or very low temperatures are used.

Consider a noise resistor delivering some noise power P_n to a load resistor of equal resistance (for maximum power transfer), as shown in Figure 8.3. Using the voltage in (8.6), the noise power in bandwidth B delivered to the load resistor is calculated as

$$P_n = i^2 R \tag{8.7a}$$

$$P_n = \left(\frac{\overline{v}}{2R}\right)^2 R \tag{8.7b}$$

8.1 Introduction to Noise

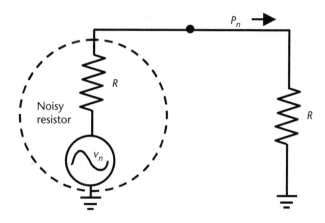

Figure 8.3 Lumped-element noisy resistor circuit.

$$P_n = kTB \quad (8.7c)$$

Solving (8.7c) for P_n at room temperature (typically accepted to be $T = 290K$) gives $P_n = 400.2 \times 10^{-23}$ W in a 1-Hz bandwidth. Placing this into more useful units gives $P_n = -174$ dBm in a 1-Hz bandwidth, or $P_n = -174$ dBm/Hz. This is theoretically the lowest possible noise level of any system at room temperature because this value is based solely upon kinetic energy due to thermal agitation of the molecules that make up matter. Note also that (8.7c) is completely independent of frequency. Thermal noise power is dependent only on temperature and bandwidth.

8.1.2.2 Shot Noise

Shot noise, also known as Schottky noise (because Schottky described it mathematically in 1928 [5]), is noise due to random fluctuations of charge carriers across a potential barrier in electronic devices. Typically, electrical current charge carriers are electrons, which can be considered as moving in a flow on a microscopic level. Because electrical current flow can be considered to comprise discrete particles, there is some random fluctuation in their movement through an electronic device. The power spectral density of shot noise is approximately broadband and flat, as shown in Figure 8.2(b). The word "approximately" is used because there is a roll off of this type of noise at approximately 10^{15} Hz because the charge carriers have a finite travel time within the device. For the frequencies of interest in RF and SOC testing (<20 GHz), the roll off is negligible.

8.1.2.3 1/f Noise

1/f (flicker noise, "pink" noise) is found in many electronic devices, and its origin is not fully understood. It was discovered soon after the invention of the transistor when scientists were trying to eliminate noise from audio transistors. It is believed that 1/f noise arises from inherent defects in the substrates of semiconductors, and unfortunately it cannot be eliminated. Some devices have exhibited 1/f noise down to frequencies much less than 1 Hz, where the noise merges with the natural drift of the device. Flicker noise measurements at these low frequencies are very difficult because of the long measurement times required.

Flicker noise has a power spectral density that varies as the inverse of frequency

$$P_{flicker} \propto \frac{1}{f} \tag{8.8}$$

In Figure 8.2(c), it is apparent that the contributions of $1/f$ noise are most significant at low frequencies, and its effects become negligible at high frequencies. Unfortunately, based on this definition, the noise power density goes to infinity as frequency goes to zero. For that reason, this definition is not valid at dc; nor have any experimental observations traced $1/f$ noise to such low frequencies. A practical description would consider $1/f$ noise to have linear power density over a given bandwidth for lower frequencies. As frequencies become higher, flicker noise levels off to a flat power spectral density like that of thermal or white noise.

8.1.2.4 Quantization Noise

Quantization noise is noise arising from the difference between the true analog value and the quantized versions of a signal in analog to digital converters (often used in test equipment). This noise is dominant in systems that have a limited dynamic range. The effects of quantization noise can be minimized or made negligible by careful design of the test system.

8.1.2.5 Quantum Noise

Quantum noise is broadband noise that results from the quantized nature of charge carriers. It is typically only seen in systems that are either cold (near 0K) or operating at very high bandwidth (greater than 10^{15} Hz); therefore, it is of little concern in production testing.

8.1.2.6 Plasma Noise

Plasma noise is noise due to the random motion of charges in an ionized gas. Gases are those such as plasma or the ionosphere. Ionization can occur in production-test equipment, even though the items listed here seem unlikely. A sparking electrical contact could create locally ionized air (plasma). A likely place that this could occur is at the contactor when testing power amplifiers. However, in general, plasma noise is not of concern when performing production test.

8.1.3 Noise Floor

The concept of the noise floor is important to understand. It is important to know the noise floor of the test equipment with which noise measurements are being made. Noise floor is the level of power below which a desired signal cannot be detected. At this minimum power level, a desired signal is said to "fall through the noise floor." Figure 8.4 shows two signals. One of the signals is above the noise floor

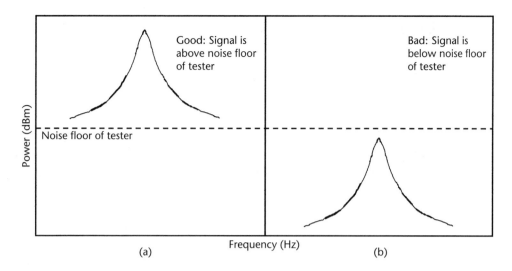

Figure 8.4 Noise floor.

and one is below. In the case of Figure 8.4(b), no evidence of the desired signal would ever be seen. This is why understanding the noise floor of the measurement instrumentation is critical.

In practice, measuring a device with high gain, for example, 20 dB (when using a spectrum analyzer to perform noise measurements), would be no problem as the gain of the device would allow the noise signal to rise through the noise floor of a spectrum analyzer. However, for devices with low gain, additional components have to be added to the noise measurement setup. Typically, a low-noise, high-gain amplifier is added in the path between the DUT and the test equipment.

Because noise is random, when measuring noise signals from a DUT, the internal noise of the test equipment must be lower than the noise being measured. With tuned receiver-based measurement equipment, reducing the resolution (or IF) bandwidth will reduce the amount of noise in the measurement system. When doing this, however, the time to perform the measurement increases, so, as with many measurements, trade-offs must be made, especially at low power-level measurements near the noise floor. When the noise power level of the DUT approaches the noise floor of the test equipment, errors will be introduced into the measurement. In a worst-case scenario, when the DUT noise power is equal to the noise floor power of the tester, the measurement will appear to be 3 dB above the noise floor (i.e., 3 dB in error) [2].

8.2 Noise Figure

8.2.1 Noise-Figure Definition

Sensitivity (often synonymous with signal-to-noise ratio) in wireless device receiver front ends is extremely important since it enables the detection and resolution of weak signals (levels down to −90 dBm and lower) commonly used in wireless LAN and personal area network (PAN) systems. In RF systems, ranging from wireless communications all the way to radio astronomy, the term *noise factor*, or F, has been defined to quantify the impact a device has on the signal-to-noise ratio:

$$F = \left. \frac{S_i/N_i}{S_o/N_o} \right|_{T=T_0=290K} \quad (8.9)$$

Equation (8.9) states that noise factor is the ratio of input signal-to-noise ratio to output signal-to-noise ratio at $T = T_0$, commonly accepted to be 290K (room temperature) [6]. In words, noise factor is the degradation of the signal-to-noise ratio at T_0. It is well known, however, that the magnitude of degradation is difficult to measure directly. Figure 8.5 depicts (8.9) showing the input power level of an amplifier (DUT) and the increased noise at the output of the amplifier resulting in a decreased signal-to-noise ratio. Note that the signal power is higher at the amplifier's output than that of the signal before entering the amplifier. However, since the amplifier adds noise (via the mechanisms described earlier in this chapter), the noise floor at the output is raised significantly. Thus, the signal-to-noise ratio at the output is less than that of the input.

The figure of merit, noise figure, or *NF*, is used more readily throughout the industry. Noise figure is simply the noise factor in units of decibels:

$$NF = 10 \log(F) \quad (8.10)$$

Many engineers use these terms interchangeably, or incorrectly in speech, which is typically not a problem as understanding is inferred from the context of the discussion. However, inadvertently mixing these two terms up in calculations can have adverse effects. Keep in mind that a perfect-noise DUT with no noise added would have a noise factor of $F = 1$ and a noise figure of $NF = 0$ dB. Thus, the potential values for noise factor and noise figure are

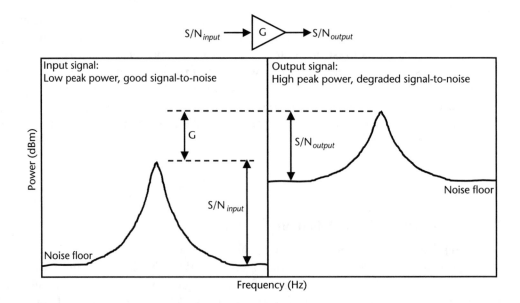

Figure 8.5 Depiction of signal-to-noise degradation.

8.2 Noise Figure

$$F \geq 1 \tag{8.11}$$

and

$$0 \leq NF < \infty \tag{8.12}$$

In a receiver front end, the LNA is the most critical stage with respect to noise figure. From the Friis equation [7]

$$F_{sys} = F_1 + \frac{F_2 - 1}{G_1} + \frac{F_3 - 1}{G_1 G_2} + \ldots + \frac{F_n - 1}{G_1 G_2 \ldots G_{n-1}} \tag{8.13}$$

it can be seen that the noise-figure performance of the first stage of a cascade is the most significant (the subscripts denote the stages). System designers try to design the first, or preamp, stage of a receiver such that the noise figure is low and the gain is high. With the high gain, G_1, the large value carries through in the denominator of each of the subsequent terms in (8.13), making their contribution to overall noise figure less significant, but not insignificant. Typically, LNAs of wireless receivers have noise figures of 1.5 dB or better. One may contest that the LNA is not the first element of a receiver chain. It is often the antenna or a bandpass filter. That is true, and the antenna and filter do contribute to the overall noise figure. However, they are passive, lossy devices that offer no gain. The LNA is the first component in the chain that offers gain.

Building upon (8.9), at $T = 290K$ (the accepted temperature of usage of wireless devices), a direct correlation exists between receiver sensitivity and noise figure:

$$NF = 10 \log\left(\frac{S_i/N_i}{S_o/N_o}\right) \tag{8.14a}$$

$$NF = 10 \log(S_i/N_i) - 10 \log(S_o/N_o) \tag{8.14b}$$

$$NF = \Delta(S_i/N_i)\big|_{dB} \tag{8.14c}$$

In other words, (8.14c) states that a 1-dB noise-figure reduction provides a 1-dB increase of receiver sensitivity gain and vice versa [6]. This should make it apparent why noise figure is such a critical parameter.

Noise-figure measurements inherently involve the characterization of low-level signals. This requires extra attention to the details of the test setup to make accurate measurements. However, in production testing compromises often need to be made as problems, such as impedance mismatch due to DUT-to-DUT impedance variations, can arise. This requires the engineer to understand all of the facts that come into play. The most important item to consider is that the noise of the equipment performing the measurement must be significantly lower than the noise that is being measured.

8.2.2 Noise Power Density

Often noise is expressed in the form of noise power density or power spectral density, expressed in units of dBm/Hz. It is therefore essential to understand this quantity and to understand how to calculate noise power from it. Understanding that noise power is specified in a bandwidth is the key to understanding why this convention is used. While many engineers do not refer to it by its formal terminology, noise power density, it is commonly inferred.

8.2.3 Noise Sources

A noise source is a one-port device that provides a known amount of noise to a DUT so that the noise figure can be calculated. The simplest (and traditional) noise source is a resistor held at a fixed temperature. The electrons within the resistor have random motion that provides kinetic energy proportional to temperature. The energy is translated into a random voltage signal having a zero average value, but a nonzero rms value given by (8.4).

If more noise power than a temperature-stabilized resistor can provide is desired, active noise sources can be used. Typical active noise sources are gas-discharge tubes or avalanche diodes. Diodes are more common in a production-test equipment environment. Figure 8.6 shows some of the various available noise sources. In its on, or hot, state the avalanche breakdown mechanism of the diode produces the noise power. The 346B-style diode noise source has been available for many years and is still widely used today. Newer designs, many surface mount, are evolving to take advantage of newer technology to have small modules available for use on the production load board. As a rule of thumb, the minimum noise power level between a hot and cold noise source must differ by at least 10 dB.

8.2.4 Noise Temperature and Effective Noise Temperature

Noise power is linear with temperature; therefore, temperature is used to characterize noise. Noise temperature is defined as the temperature that a resistor would have to be placed at to have the same available noise power spectral density as the actual noise source. This definition is based upon the calculation of thermal noise power in (8.7) as

Figure 8.6 Types of noise sources. (*Courtesy of:* Noise/Com Corporation.)

$$T_a = \frac{N_a}{kB} \tag{8.15}$$

where the subscript "a" denotes "available," and N_a is the available noise power (in watts) of the actual source.

A slightly different, but more useful, value is effective noise temperature:

$$T_{ne} = \frac{N_e}{kB} \tag{8.16}$$

where N_e is the emerging power under the assumption that the power spectral density is constant across the measurement bandwidth. Effective noise temperature is calculated from the power emerging from the noise source when it is terminated in a nonreflecting and nonemitting load.

Effective noise temperature is related to noise temperature as

$$T_{ne} = T_a\left(1 - |\Gamma|^2\right) \tag{8.17}$$

where Γ is the reflection coefficient (see Chapter 4) of the one-port noise source.

While these two quantities vary only in verbal definition, it is effective noise temperature, T_{ne}, that is used in calculations surrounding noise figure in this chapter.

8.2.5 Excess Noise Ratio

Excess noise ratio (ENR) is a term used to describe the output of a noise source when it is used as an input stimulus to a circuit. The definition of ENR is

$$ENR = \frac{\text{Noise power difference between hot and cold sources}}{\text{Noise power at } T_0} \tag{8.18a}$$

$$ENR = \frac{k(T_h - T_c)B}{kT_0 B} \tag{8.18b}$$

$$ENR = \frac{T_h - T_c}{T_0} \tag{8.18c}$$

where T_h is the equivalent noise temperature of the noise source in the on, or hot, state (in kelvins), T_c is the equivalent noise temperature of the noise source in the cold state, and T_0 is the reference temperature (assumed to be the standard 290K). Most often, in production testing of DUTs, T_c is simply T_0, making (8.18c)

$$ENR = \frac{T_h}{T_0} - 1 \tag{8.19}$$

The above definitions provide ENR in linear units. In test equipment at RF frequencies, it is more common to use logarithmic values hence

$$ENR_{dB} = 10\log(ENR) \tag{8.20}$$

The ENR for typical noise sources used in production and bench top testing of wireless devices is about 15 to 20 dB.

If possible, it is desirable to use a low-ENR noise source when the noise figure is low enough to be measured with those conditions. Because a low-ENR indicates lower noise power levels, this means that the tester or measuring equipment will require minimal dynamic range and be less likely to operate in the nonlinear range. Additionally, the use of a low ENR noise source has a more constant impedance between the on and off states of the noise source. This is because a low ENR noise source is typically a high ENR noise source with an attenuator.

Noise sources are calibrated by the National Institute of Standards and Technology (NIST). So as long as the calibration is legitimate and up-to-date, there are not a lot of additional steps that one can take to improve it. Some noise-figure measuring equipment requires manual entry of ENR calibration tables. This can lead to error in measurement due to typographical errors. It is best to utilize data transfer from computer to tester if one is available for this task.

8.2.6 Y-Factor

The Y-factor is a ratio of hot to cold noise powers (in watts) and is defined as

$$Y = \frac{N_h}{N_c} \tag{8.21}$$

If the noise source is at room temperature, and the cold state is that of a noise diode simply turned off, then $T_c = T_0$, and (8.20) becomes

$$Y = \frac{N_h}{N_0} \tag{8.22}$$

Because the Y-factor is a ratio of the measurement of two power levels, absolute accuracy of the test equipment isn't the most critical issue. It is of more importance that it be repeatable so that whether the diode is on or off, the test equipment measures under the same conditions. The Y-factor is the foundation for most modern noise-figure measurements and calculations.

8.2.7 Mathematically Calculating Noise Figure

Now that a few terms have been introduced, the noise figure of a device can be calculated. Having acquired the ENR value of the noise source (usually provided by the manufacturer) and having measured the Y-factor, noise figure is simply calculated as

$$F = ENR/(Y - 1) \tag{8.23}$$

Note that the ENR value and Y in (8.23) must be in linear units. However, the ENR value of a noise source is almost always supplied in decibel format. Therefore, a more convenient calculation is

$$NF = 10\log(F) \tag{8.24a}$$

$$NF = 10\log(ENR) - 10\log(Y-1) \tag{8.24b}$$

$$NF = ENR_{dB} - 10\log(Y-1) \tag{8.24c}$$

Although upon immediate inspection it looks as if there is no temperature dependence in this calculation of noise figure, note the inherent dependence of ENR on equivalent noise temperature due to (8.18).

8.2.8 Measuring Noise Figure

There are multiple ways to acquire the various parameters needed to calculate noise figure, but typically, only three are used in practice for testing wireless and SOC devices. Those are termed the direct method, the Y-factor method, and the vector-corrected cold noise method [8]. The direct method is the simplest to implement for production, but it is limited to devices with high gain. The Y-factor method is also a relatively straightforward technique to implement for production testing and is the foundation for most noise-figure meters and analyzers. The cold noise method is a vector-corrected method that is designed for production testing; however, due to the extensive correction calculations (that make it so accurate), it may require more time to perform. Each technique has its advantages and disadvantages, and it is up to the test engineer to make a choice based on his particular needs. The three methods are discussed next.

8.2.8.1 The Direct Calculation of Noise Figure

If the DUT has a large amount of gain, such as an SOC receiver, it may be acceptable to measure the noise directly and calculate noise figure from

$$F_1 = \frac{N_0}{kT_0 BG} \tag{8.25}$$

This method can be very convenient if the DUT contains other measurements in which spurious-free frequency-domain data has been acquired. That way, a point away from the peak on the signal of interest may be taken to be the noise power, N_0. B is the measurement bandwidth. The gain, G, of the DUT must also be measured, and it is likely that this has been done at some other time in testing the DUT. Finally, kT_0 is simply the value –174 dBm/Hz. Therefore, this becomes almost a free measurement (except for calculation time).

Having acquired N_0, B, and G, the noise figure can be calculated by placing (8.25) into logarithmic values,

$$NF = N_{0|dB} - (-174\,\text{dBm/Hz}) - B_{|dB} - G_{|dB} \tag{8.26}$$

8.2.8.2 Measuring Noise Figure Using the Y-Factor Method

Figure 8.7 is a plot of output noise power (in watts) versus source temperature (in kelvins). This plot will be used as when implementing the Y-factor method for

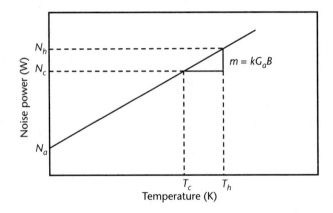

Figure 8.7 Output noise power versus temperature.

determining noise figure of any two-port device. In this description, noise factor will be calculated and then converted to noise figure at the end. When the noise powers of two significantly different noise sources (hot and cold) are plotted against their equivalent noise temperatures, a lot of information can be gathered. The slope of the line is

$$m = \frac{(N_2 - N_1)}{(T_2 - T_1)} \tag{8.27a}$$

or

$$m = \frac{\Delta N}{\Delta T} \tag{8.27b}$$

The gain (whether greater or less than one) of a DUT, G, linearly multiplies with the noise power of (8.7c) to lead to

$$N = kGBT \tag{8.28}$$

Using (8.27b) and (8.28), the slope of the line in Figure 8.7 is

$$m = kGB \tag{8.29}$$

Furthermore, the line segment made between the two points can be extrapolated to the y-intercept, which will be termed N_a, or the noise added by the DUT. Therefore, noise factor can be calculated as in (8.23).

Figure 8.8 shows a typical Y-factor noise-figure measurement setup. Note that this can be viewed as a cascade of two stages, where the DUT is the first stage and the tester (receiver) is the second stage. Taking the first two terms of (8.13) and rearranging them,

$$F_1 = F_{12} - \frac{F_2 - 1}{G_1} \tag{8.30}$$

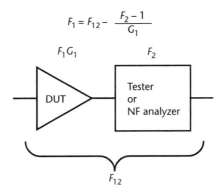

Figure 8.8 Y-factor measurement setup.

provides a means to calculate the noise factor of the DUT (F_1). Additionally, F_{12} is the overall noise factor of the DUT and tester, F_2 is the noise figure of just the tester, and G_1 is the gain of the DUT.

The first step is a calibration step, as shown in Figure 8.9(a). During this calibration step, the noise source is connected directly to the tester receiver. After the two power levels are measured, corresponding to the applied hot and cold noise sources, the Y-factor, noise factor, and gain for the tester are calculated as

$$Y_2 = \frac{N_{h2}}{N_{c2}} \tag{8.31}$$

and

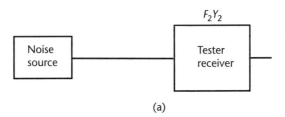

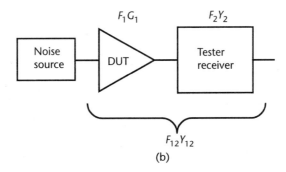

Figure 8.9 Y-factor measurement process: (a) calibration stage, and (b) measured stage.

$$F_2 = \frac{ENR}{Y_2 - 1} \qquad (8.32)$$

ENR is the excess noise ratio of the noise source in linear units.

Next, with the DUT inserted, as in Figure 8.9(b), the hot and cold power measurements are taken again to determine the Y-factor, noise factor, and gain of the DUT and tester combination:

$$Y_{12} = \frac{N_{h12}}{N_{c12}} \qquad (8.33)$$

and

$$F_{12} = \frac{ENR}{Y_{12} - 1} \qquad (8.34)$$

In both the calibration stage and the measurement stage, the measurement accuracy and repeatability can be improved by having the hot and cold noise sources repeatedly cycled on and off. By taking multiple measurements, correspondingly, this will allow for averaging to obtain a better value for the Y-factor. Also, when performing the hot and cold measurements, it is essential to ensure that the hot and cold power levels of the noise source are linear with respect to each other.

The only remaining value to determine is G_1, the gain of the DUT. That can be found from the noise power values that have already been measured:

$$G_1 = \frac{N_{h12} - N_{c12}}{N_{h2} - N_{c2}} \qquad (8.35)$$

After obtaining all of these values, they are inserted into (8.30) to solve for the DUT noise factor, F_1, and thus, from (8.24a), the noise figure is

$$NF_1 = 10 \log(F_1) \qquad (8.36)$$

Following these steps should prove to be very straightforward, but to implement the Y-factor method in a production environment, some extra steps have to be taken. As long as they are considered at the early stages and prior to load board design, all should be well. Note that the method entails performing a calibration step. This is done without the DUT in the measurement path. This is often implemented through the use of switches on the load board. They allow the noise source to be connected directly to the tester (receiver) for the calibration step, and they allow the noise source to be applied to the DUT, as in the measurement stage.

If a noise diode is going to be used, consider that many noise sources use a 28-V supply as a standard (lower-voltage noise diodes are available as well). The dc supplies that are to be used must be free of noise themselves, as well as being capable of allowing the noise source to be switched on and off, perhaps in a cyclic fashion, if multiple noise power measurements are performed for added averaging as mentioned above.

8.2.8.3 Measuring Noise Figure Using the Cold Noise Method

The Y-factor method of measuring noise figure is the most accurate method as long as the DUT is perfectly impedance matched to the tester. Due to impedance variation from DUT to DUT, it is nearly impossible to have a perfect match for all DUTs, even those within the same tested lot. Because the Y-factor method uses only scalar measurements, it does not take into account the phase information that can be used to correct for the impedance mismatch.

The cold noise method of noise figure has been created to account for the impedance mismatch between the DUT and tester. It is based upon a full S-parameter measurement (magnitude and phase of all four S-parameters for a two-port device, for example). The method is similar to the Y-factor method, but has the added advantage of using a correction algorithm to correct for mismatches between the DUT and tester. These algorithms are very computationally intensive, but with today's high-speed processors, should add little to no overhead to the test time.

The primary difference between this method and the Y-factor method is that in this method, the noise factors referred to in (8.30) are functions of the reflection coefficient, Γ (see Chapter 4), such that

$$F_1(\Gamma) = F_{12}(\Gamma) - \frac{F_2(\Gamma) - 1}{G_1} \tag{8.37}$$

The two primary steps of the cold noise technique are as follows:

1. The calibration process is performed, similar to that of the Y-factor method, with the difference being that the measurements are full S-parameter measurements. From this, correction factors for impedance mismatching are created.
2. With the DUT inserted (or switched in), full S-parameter measurements are made to find the true available gain of the DUT, rather than the "insertion" gain, as found from the Y-factor method. The available gain is used in conjunction with noise power and placed into (8.25). The name cold noise arises because the only noise source at measurement time is a 50-ohm termination at the input of the DUT.

Refer to [8] for more detailed calculations of the cold noise technique. The information provided here should be enough to help a test engineer decide which method is best for a particular application. In general, if the DUT is well impedance matched to the tester (and has minimal match variation between DUTs), then the Y-factor would be the best choice based on simplicity. If there is a poor match between the DUT and tester or a lot of variation between DUTs (such that a perfect matching network that meets all needs cannot be created on the load board), then the cold noise technique is more suitable.

8.2.9 Noise-Figure Measurements on Frequency Translating Devices

Up to now, most of this discussion has focused on measuring the noise figure of amplifiers or two-port devices. When measuring the noise figure of frequency-translating devices, such as mixers, there are some differences in behavior that need to be addressed. One primary difference is that when measuring the noise figure of mixers, the noise source ENR is that of the microwave frequency, but the input of the measurement instrument is tuned to the IF frequency of the device. To assure that this is not a problem, it is necessary to have a broadband noise source that extends between the RF and IF frequencies, or more importantly, one that has the same ENR at both frequencies.

Many mixers are passive and have loss (conversion loss) associated with them. This loss corresponds to the value G in (8.26). Placing a linear gain value of less than one (corresponding to loss) into (8.13), the Friis equation, will show that the second stage of a cascade system can have a large impact on the overall noise figure. If a mixer of this type is being used in the tester or noise-figure analyzer (for example, as a downconverter to a system IF frequency), then it can introduce significant measurement error.

If lossy mixers are the DUT being tested, the Y-factors may be very small. To remedy this, it is recommended that you use a noise source with a high ENR, for example, higher than 17 dB.

There are two principle types of mixers, single sideband (SSB) and double sideband (DSB). Noise figures can be measured for both; however, care must be taken and an understanding of the effects of the various noise power levels is important, as is interpreting the results. If measuring with a noise-figure analyzer, actual conditions are measured. That is, if the mixer rejects one sideband, a SSB result is displayed. Similarly, if the mixer converts both sidebands, DSB results are displayed. Thus, care should be taken when interpreting results, since confusion can occur if DSB results are used to predict performance of an SSB system. For example, if DSB results are taken for an SSB mixer, the noise figure will be 3 dB lower than it is in reality. This could potentially cause problems in that the mixer's noise figure is really 3 dB higher than tested. Additionally, measured gain will be 3 dB higher for DSB measurements of SSB mixers because the measured bandwidth is twice the calibrated value.

8.2.10 Calculating Error in Noise-Figure Measurements

Based on (8.13), the error that is introduced in noise-figure measurements can be piecewise determined from [8]

$$\Delta NF = \sqrt{\left[\frac{F_{12}}{F_1}\Delta NF_{12}\right]^2 + \left[\frac{F_2}{F_1 G_1}\Delta NF_2\right]^2 + \left[\frac{F_2-1}{F_1 G_1}\Delta G_1(\text{dB})\right]^2 + \left[\left(\frac{F_{12}}{F_1} - \frac{F_2}{F_1 G_1}\right)\Delta ENR\right]^2}$$

(8.38)

The detailed derivation of this equation is given in [9, 10].

The first term, consisting of ΔNF_{12}, accounts for mismatch between the noise source and the DUT and the overall instrument uncertainty. The instrument error is most often small as long as the user has chosen the best-fit test equipment for the given DUT.

The second term represents error due to the tester noise figure. For example, accuracy and repeatability are lost when the tester has a high noise figure relative to the DUT. If a preamplifier is not added to the tester, then this term can be a significant contributor.

The third term is dependent on the DUT gain. If the DUT gain is high, then note that the large number, G_1, in the denominator of the third term reduces the overall effect of ΔG_1. If, however, G_1 is small, then ΔG_1 becomes more significant.

Finally, the fourth term accounts for any uncertainty or error due to the noise source, or ENR. This can range from bad calibration data to the impact on accuracy due to the cold noise temperature's being different from the assumed 290K.

8.2.11 Equipment Error

When making noise-figure measurements, it is important to be aware of the equipment, or tester, and the methods that are used to perform the measurement. If, for example, the tester employs a downconversion scheme in making the power measurements, then it is necessary to know whether a double-sideband or single-sideband mixer is used internally to the tester. The power measured in the unwanted sideband is measured and will add erroneously into the overall power measured.

If the tester has a high noise figure itself, this limits the accuracy and repeatability with which the noise figure can be measured. It is common practice to add a low-noise, high-gain preamplifier to the input of the tester. This preamplifier then becomes part of the tester and enables the tester noise figure to be reduced as it becomes the first stage in (8.13).

Nonlinearity is a problem in both the measurement equipment, as well as in the noise source. Any nonlinear effects within the detector will reappear in every calibration and measurement. Nothing one does to the DUT or external environment will change this. To minimize nonlinear effects—for example, if the device noise figure to be measured is quite low—then it is recommended to use a low ENR source. The low ENR will require less dynamic range of the detector, hence keeping the instrument in a linear mode of operation.

8.2.12 Mismatch Error

Impedance mismatch between noise source and DUT and DUT and tester is perhaps the largest contributing source of error. As explained earlier, from an accuracy standpoint it warrants the use of full S-parameter based noise-figure measurement. The two primary problems arising due to mismatch are that noise power is lost at an interface when mismatch is present and reflections of the noise power signal give rise to unpredictable effects.

The noise source can impact mismatch error. Low ENR sources with high internal attenuation are a best choice due to the lower VSWR and greater consistency of match between on and off impedances. Measurement of noise figure on DUTs with high gain are less susceptible to the effects of mismatch since higher gain reduces the relative contribution of the second-stage noise-figure component from the instrument [see (8.13)].

If S-parameter-based measurements of noise figure are properly made, the errors associated with mismatch can be reduced significantly. However, it should be noted that this method could be computationally cumbersome. Because a full vector-based measurement of the DUT is performed, error correction terms can be applied to the noise-figure measurement and provide a result similar to that if the DUT had been in a perfectly matched environment.

8.2.13 Production-Test Fixturing

When measuring noise figure in a production environment, there are even more sources of error. For production testing of packaged DUTs, a test fixture, or contactor, is used on a load board with a fixed matching network. Due to variation between DUTs, the match between the DUT and the load board (and ultimately, the tester) will vary from DUT to DUT. For production noise-figure measurement of wafers, a wafer probe is used. In either case, the means of contacting the DUT will introduce error. It will add loss and mismatch. Ideally, the measured output power of the DUT has to be corrected, and the effect of the fixture or probes has to be removed.

A production-test fixture is a common place to look for noise being introduced into the system. The most difficult problem with shielding the production-test fixture is finding a shielding means that can physically fit within the constraints of the DUT handler that is used. While it is nearly impossible to completely remove the fixturing effects from the noise-figure measurement, a typical technique is to use scalar correction that can compensate for the loss at the input of the device.

8.2.14 External Interfering Signals

With the proliferation of mobile phones, pagers, and the like in the vicinity of the test environment, unwanted interfering signals can degrade the performance of the noise-figure measurement. It is not uncommon for wireless LAN networks or microwave ovens to produce interfering signals.

From an electronics standpoint, it is common practice to locate an interference source and shield it to remove the cause of the interference. In the case of production noise-figure measurements, it is not uncommon to place an entire test system within a screen room or a type of Faraday cage structure. As a rule of thumb, shielding should reduce extraneous signal levels by 70 to 80 dB.

Noise due to the measurement instrument itself is not usually a problem, as commercially available instrumentation is typically well shielded. However, be aware that an older computer integrated into the test setup can add noise as shielding requirements were less stringent years ago.

8.2.15 Averaging and Bandwidth Considerations

Finally, a word must be said about the residual jitter that is present simply due to the fact that noise is a random electrical signal. Repeatability errors will be introduced because the measurement is performed over a finite time (infinite time is required to acquire the true value of noise, but that is obviously not practical). Measurement averaging should be used when possible. Of course, measurement averaging adds test time, but through observation, a compromise must be determined between the amount of averaging and the quality of the repeatability desired. The general relationship of jitter in the signal, resulting from averaging, is

$$\text{Residual jitter} \propto \frac{1}{\sqrt{N}} \tag{8.39}$$

where N is the number of averages.

Alternatively, if the bandwidth of the measured noise is wide enough, enough noise data may be collected. The relationship between bandwidth B and residual jitter is

$$\text{Residual jitter} \propto \frac{1}{\sqrt{B}} \tag{8.40}$$

This means that either the noise measurement should be performed over a large bandwidth or with many measurement averages to obtain the best repeatability.

8.3 Phase Noise

8.3.1 Introduction

Phase noise is a parameter that measures the spectral purity of a signal. It is particularly referenced to a sinusoidal, or CW, waveform. It is associated with the term *jitter*. The principle difference between the two is that jitter is a property described best by relating to the time-domain, while phase noise is best described when related to the frequency domain.

A pure sine wave in the frequency domain will look like an impulse function with all of the energy concentrated at exactly the carrier frequency. In reality, if the frequency space around the carrier frequency is explored, there will be energy located at the adjacent frequencies. This energy is due mostly to phase noise. Its behavior is that of $1/f$ noise.

In practice, phase noise is represented in the frequency domain as shown in Figure 8.10. It measures the spectrum of phase deviation. This is its most common representation, as a single-sideband power measurement in a 1-Hz bandwidth at

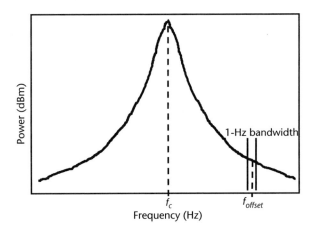

Figure 8.10 Typical phase-noise specification.

some frequency away from, but relative to, a carrier power. For example, a phase-noise specification for a VCO on an SOC device might be stated as

$$\text{Phase noise} = -90 \text{ dBc/Hz at 10-kHz offset}$$

which means that the measured power in a 1-Hz bandwidth at 10 kHz away from a carrier signal is 90 dB lower than the power of the carrier signal.

In wireless digital communications, the modulated signals contain information that is determined by the phase state of the signal. If the signal encounters too much phase noise, the relative and absolute positions of the information upon demodulation will be disturbed, and the information will be unable to be extracted. Figure 8.11 shows an I/Q constellation for a digitally modulated signal. The radial errors are due to amplitude noise, while the rotational errors are due to phase noise.

The measurement of phase noise can also help to indicate other items, such as bit error rate (BER) and signal spreading. For example, in a GSM system, if phase noise is measured at a 200-kHz offset, the resultant value will tell how much energy is falling into the adjacent channel, as the 200-kHz offset is exactly the position of the adjacent channel. Any contributions from one channel into another channel can introduce such impairments as deteriorated BER [11].

8.3.2 Phase-Noise Definition

A pure sine wave is typically represented by the following equation:

$$v(t) = V_0 \sin 2\pi f_0 t \tag{8.41}$$

where V_0 is the peak voltage amplitude of the signal, and f_0 is the carrier frequency.

The noise that can occur on this signal can exist in the form of amplitude noise, phase noise, or both. If the sine wave exhibits noise in the form of both amplitude and frequency, then (8.41) changes to

$$v(t) = [V_0 + a(t)] \sin[2\pi f_0 t + \phi(t)] \tag{8.42}$$

where $a(t)$ is the amplitude noise, and $\phi(t)$ is phase noise. The noise introduced by $a(t)$ and $\phi(t)$ is shown in Figure 8.12 in both the time and frequency domains. Notice the spreading due to phase noise, as well as the amplitude modulation sidebands.

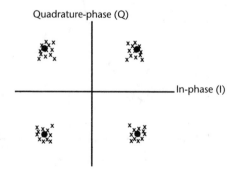

Figure 8.11 Digital modulation I/Q constellation impaired by amplitude and phase noise.

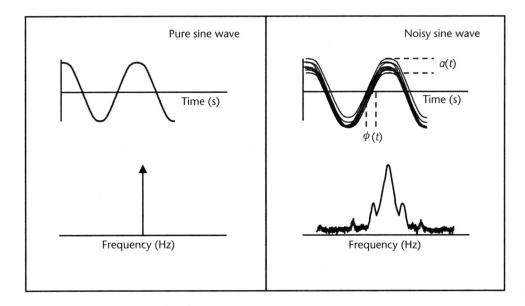

Figure 8.12 Noise on a sinusoidal waveform.

The definition given in (8.42) is also that of a signal having amplitude and phase modulation. This is actually the case. The amplitude and phase modulating signals are those of random noise processes. Phase variations are caused by random processes giving rise to thermal fluctuations that modulate the pure signal. Because phase and frequency are directly related, the variations can be consolidated and grouped under the category of "phase." Just as with a phase-modulated signal, when viewing the noisy signal in the frequency domain, sidebands, due to the noise, arise.

For this discussion of phase noise, it will be assumed that the amplitude-noise contribution is much less than that of the phase-noise contribution.

The phase-noise contribution, $\phi(t)$, could include both long- and short-term phase variation. In general, the long-term variation is considered frequency drift, while the term *phase noise* is reserved for the short-term variation.

Phase noise is the Fourier transform (or power spectral density) of the phase component of a sinusoidal signal scaled to dBc/Hz, where dBc means the power relative to the overall carrier power. Phase noise in the frequency domain can be expressed as

$$\mathcal{L}(f) = \frac{P_{offset}}{P_{carrier}} \quad (8.43)$$

where P_{offset} (watts) is the rms noise power in a 1-Hz bandwidth at a frequency f Hz away from the carrier, and $P_{carrier}$ (watts) is the rms power of the carrier. The units of (8.43) are dimensionless.

The $\mathcal{L}(f)$ symbol is termed the *Laplacian* and it represents the frequency notation of phase noise. It is used almost universally throughout the literature. Often the units of $\mathcal{L}(f)$ are expressed in decibels, making (8.43) become

$$\mathcal{L}(f) = P_{offset}(\text{dBm/Hz}) - P_{carrier}(\text{dBm}) \quad (8.44)$$

The units of (8.44) are dBc/Hz.

As the frequency offset approaches zero, the variation is more appropriately termed *frequency drift*. Near-in phase-noise measurements often pose the difficult task of requiring equipment to measure both the carrier power and the phase noise. On very high stability oscillators (DUTs), these measurements can push the limits of the dynamic range of the test equipment.

In the frequency domain, a pure sine wave would be represented as an impulse waveform or an infinitely narrow peak. In practice, there is always some sideband present. This is inherently due to fundamental physics. From a physics standpoint, referring to both the fundamentals of the Fourier transform and the Heisenberg Uncertainty Principle, the only way to have an infinitely narrow peak would be to measure the signal over the time period of −infinity to +infinity, and that is obviously not possible (recall the topic of this book, production testing). It is obvious that trade-offs have to be made in general phase-noise measurements and, then, potentially further in production phase-noise measurements to allow for cost-effective analysis.

8.3.3 Spectral Density–Based Definition of Phase Noise

Another method of defining phase noise is based upon the one-sided power spectral density. Reference [2] defines this based on the random nature of the phase instabilities. Using the concept of phase noise being equivalent to phase modulation by a noise source, the spectral density is defined as

$$S_\phi(f) = \phi^2(f)\frac{1}{B} \tag{8.45}$$

where B is bandwidth (in hertz) and the units of $S_\phi(f)$ are radians^2/Hz.

Recall from (8.43) that power spectral density describes the power distribution as a continuous function expressed in units of energy within a given bandwidth. The short-term instability is measured as low-level phase modulation of the carrier and is equivalent to phase modulation by a noise source.

The traditional definition of phase noise, as in (8.44), is the ratio of the power in one phase modulation sideband per hertz to the total signal power, usually expressed in decibels relative to the carrier power per hertz of bandwidth (dBc/Hz).

This traditional definition may be confusing when the phase variations exceed small values because it is possible to have spectral density values that are greater than 0 dB, even though the power in the modulation sideband is not greater than the carrier power.

IEEE Standard 1139 [12] has been modified to define phase noise as

$$\mathcal{L}(f) = S_\phi(f)/2 \tag{8.46}$$

to eliminate any confusion [13].

8.3.4 Phase Jitter

Using the spectral density–based definition of phase noise, phase jitter [$\phi^2(f)$], defined as the total rms phase deviation within a specified bandwidth, is calculated as

$$\phi^2(f)\big|_{radians} = \int_{f_1}^{f_2} S_\phi(f)df \qquad (8.47)$$

or

$$\phi^2(f)\big|_{degrees} = \frac{360}{2\pi}\int_{f_1}^{f_2} S_\phi(f)df \qquad (8.48)$$

Units of decibels may be used when phase jitter is relative to 1 radian (rms). Additionally, it is not possible to obtain $S_\phi(f)$ from phase jitter unless the shape of $S_\phi(f)$ is known.

8.3.5 Thermal Effects on Phase Noise

Thermal noise can limit the extent to which phase noise can be measured. From the fundamental description of thermal noise, described by kTB, at room temperature (290K) noise power is –174 dBm/Hz. Because phase noise and amplitude noise are uncorrelated [see (8.42)], each contributes equiprobably to kTB. The phase-noise power contribution to kTB is –177 dBm/Hz, and the AM modulation noise power contributes –177 dBm/Hz (note that each is 3 dB less than the total thermal power).

8.3.6 Low-Power Phase-Noise Measurement

Measuring phase noise of low-power signals can be difficult. However, a low noise amplifier can be used to boost the device carrier power signal to levels necessary for successful measurements, but the theoretical phase-noise measurement is limited by the noise figure of the amplifier and the low signal power from the signal to be measured:

$$\mathcal{L}(f) = -177(\text{dBm/Hz}) + NF(\text{dB}) - P_{DUT}(\text{dBm}) \qquad (8.49)$$

where –177 dBm/Hz is the theoretical noise power due to phase noise at room temperature, NF is the noise figure of the amplifier, and P_{DUT} is the power of the signal from the DUT before it is amplified.

8.3.7 High-Power Phase-Noise Measurement

At high power levels (i.e., above 0 dBm) attenuators are often used in test equipment receivers. These reduce the signal-to-noise ratio and deter measurement. Therefore, it is recommended to perform phase-noise measurements at lower power levels.

8.3.8 Trade-offs When Making Phase-Noise Measurements

There are two trade-offs when making phase-noise measurements:

1. Measurement speed versus measurement information;
2. Measurement ease versus measurement sensitivity.

Obviously, for production testing, it is desirable to have a fast and easy-to-perform measurement; however, there is always a different median point that must be found for each application.

A fast measurement will provide reduced test times, but at the expense of providing little information about the signal. If more detailed information needed, it comes at the expense of longer test times, based solely on the fact that more data from the device is needed.

Since phase-noise measurement information is a spectral distribution of noise data, the foremost contributor to measurement speed is the offset frequency selected since this determines the longest time record or the narrowest resolution bandwidth. If the measurement equipment is using averaging (which is most often the case), it has the next highest contribution to measurement time. If using a spectrum analyzer or spectrum analyzer–based equipment, then for offset frequencies that are far from the carrier, the resolution bandwidth can be much larger than the resolution bandwidth for offset frequencies that are near the carrier. The narrower the resolution bandwidth, the more time required to gather the measured data.

An easy-to-perform measurement is often synonymous with a quick and custom measurement tailored to a specific device. This usually means that reconfiguring the measurement setup for another different device takes more of an effort. Often with measurement setups, it is desirable to have a setup that can meet the needs of multiple devices. Particularly for low-power-signal phase-noise measurement, if that type of sensitivity is needed, it often comes at the expense of difficult or expensive measurement setups.

8.3.9 Making Phase-Noise Measurements

There are two critical items to be aware of when measuring phase noise or designing phase-noise measuring equipment:

1. The measuring receiver must have a lower noise floor than the signal to be measured.
2. Any local oscillator in the measuring receiver must have better phase noise than that of the signal to be measured.

In order for a measurement receiver (or spectrum analyzer) to be able to measure a device's phase noise, it is imperative that the noise floor of the receiver be low enough that it is not higher than the phase noise to be measured. Figure 8.13(a) shows a legitimate measurement setup where the signal can easily be discerned from the noise floor of the receiver. Note that the lower-powered signal, Figure 8.13(b), has its desired phase-noise measurement point below the noise floor of the measurement receiver. If it turns out that the measurement is just of the noise floor of the receiver, then one solution is to increase the carrier power of the device to have the desired measured signal overcome the receiver noise floor.

Another concern is that the local oscillator in the measuring instrument's receiver must not contribute phase noise that will deter the measurement. In almost any receiver, the input signal (signal to be measured) will be mixed with the measuring instrument's local oscillator to produce a new (IF) frequency that is analyzed. If phase noise from the receiver's local oscillator is introduced, it may be interpreted as

8.3 Phase Noise

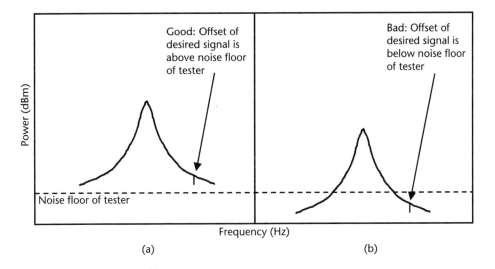

Figure 8.13 Effects of the noise floor of the receiver: (a) good, and (b) bad.

that of the device. If it is significant and at the frequency of interest, it will introduce measurement error. Assuming that the receiver has a spectrally clean LO, this only becomes an issue during attempts to measure near-in phase noise. Unfortunately, low-phase-noise RF sources (LO sources) often come at the expense of being slower.

If the measured values are higher than expected, and there is suspicion that the phase noise measured is being limited by the measurement system, then remove the device and take a raw phase-noise measurement. That is the phase noise of the receiver, and it can be used as an indication of how well a device's phase noise can be measured. Figure 8.14 demonstrates the effect of the phase noise of the receiver (tester). The dotted line is the phase noise of the receiver. In Figure 8.14(a), the phase noise of the receiver falls beneath the phase noise of the device to be measured (solid line). Thus, the receiver is not limiting the ability to measure the phase noise of

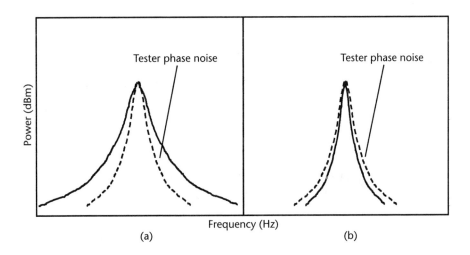

Figure 8.14 Phase noise of the measurement system. (a) Good: proper phase noise measurement, and (b) Bad: the phase noise of the receiver hides the signal of the DUT.

the DUT. In Figure 8.14(b), the phase noise of the receiver hides the signal of the device. The case in Figure 8.14(b) should be avoided as it prohibits measurement of the true phase noise of the DUT.

Someimes, phase noise cannot be directly measured on the device. Methods such as external quadrature mixing and clean amplification of the device's signal via an external LNA are required.

8.3.10 Measuring Phase Noise with a Spectrum Analyzer

The easiest method, as well as the traditional method, of measuring phase noise is to use a spectrum analyzer. However, it must be noted that when using a spectrum analyzer, the measurement is of noise in general. It is not limited to just phase noise. Both amplitude and phase-noise contributions are taken into account when using the spectrum analyzer.

Referring to (8.42), the definition of the measurement of phase noise is the Fourier transform of only $\phi(t)$. However, a spectrum analyzer provides the Fourier transform of the entire waveform $v(t)$. If the proper conditions are met, then an accurate measure of phase noise can be obtained with a spectrum analyzer.

The universal assumption when measuring phase noise with a spectrum analyzer is that phase noise is the dominant noise present. It is also a condition that the phase noise must be relatively good, or low-level. Thus, from (8.42),

$$|a(t)/V_0| << |\phi(t)/2\pi| << 1 \qquad (8.50)$$

If this is the case (as is typically assumed in practice), then the calculation of phase noise from a spectrum analyzer (or any test equipment that uses an IF filter, similar to a spectrum analyzer) is

$$\mathcal{L}(f) = P_{offset} - P_{carrier} - 10\log(B) \qquad (8.51)$$

where P_{offset} and $P_{carrier}$ are the power read from the network analyzer display. B is the resolution bandwidth (or IF filter) setting in hertz, and its term is subtracted to normalize to 1 Hz.

8.3.11 Phase-Noise Measurement Example

Measurement of phase noise in practice often leads to incorrect measurements based on the engineer having a misconstrued interpretation of the definition. As an example, a phase-noise measurement will be made using a simple, very spectrally pure, signal-generator output sent to a spectrum analyzer.[1] Figure 8.15(a) shows a spectrum analyzer display of the signal generator output. The spectrum analyzer resolution bandwidth (RBW) is set to 100 Hz and marker 1 (2 kHz away from the carrier) is 67.86 dB lower than the carrier (marker 1R). From (8.51) the phase noise is calculated as:

---

1. C. C. Kang, Agilent Technologies, personal communication to author, July 2003.

8.3 Phase Noise

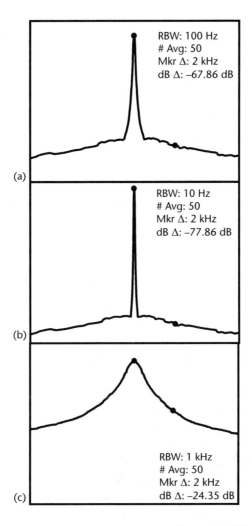

Figure 8.15 Phase-noise measurement on a spectrum analyzer with different hardware settings.

$$\mathcal{L}(f) = -67.86 - 10\log(100) = -87.86 \, \text{dBc/Hz} \tag{8.52}$$

The 10 log(100) term normalizes the 100-Hz resolution bandwidth to 1 Hz to be consistent with the definition of phase noise. The appropriate specification of this measurement is, "the phase noise is 87.86 dBc/Hz at 2-kHz offset from the carrier."

In Figure 8.15(b) the same signal has been applied to the spectrum analyzer, but the resolution bandwidth of the spectrum analyzer has been reduced to 10 Hz. In this case the difference in the carrier and the signal at a 2-kHz offset is −77.86 dB. The narrower resolution bandwidth has seemingly provided lower phase noise. However, while the level of the display has been reduced, so has the normalization factor; hence, the calculated phase noise is once again

$$\mathcal{L}(f) = -77.86 - 10\log(10) = -87.86 \, \text{dBc/Hz} \tag{8.53}$$

Reducing the spectrum analyzer resolution bandwidth will never interfere with making a proper phase-noise measurement; however, it will increase the measurement time. Note, though, that problems may occur when one attempts to increase the resolution bandwidth.

Once again, the same signal is applied to the spectrum analyzer in Figure 8.15(c). In this case the difference in the carrier and the signal at a 2-kHz offset is –24.35 dB. Calculating phase noise,

$$\mathcal{L}(f) = -24.35 - 10\log(1,000) = -54.35 \text{ dBc/Hz} \qquad (8.54)$$

provides an unexpected, different result, which is a consequence of the fact that the chosen resolution bandwidth is too wide and the phase noise is hidden beneath the IF filter skirt of the network analyzer.

8.3.12 Phase Noise of Fast-Switching RF Signal Sources

Although it is not apparent in the laboratory on bench top RF signal sources, almost always, the quality of the phase noise of the RF signal source is inversely proportional to the amount of time it takes to change frequencies or power levels. Often, with production-test systems or rack-and-stack hardware architecture, designers of the hardware aim to find RF sources that exhibit the fastest frequency- and power-switching speeds possible. It is a logical assumption to do this, and for many devices, it is the appropriate choice. However, modern wireless and SOC devices require higher performance and tighter tolerances in the area of phase noise. It is often the case that the available fast RF signal sources of a test system are inadequate to perform these stringent measurements, and this results in measuring the phase noise of the signal source, as is shown in Figure 8.14(b). When choosing a test system, the availability of a low-phase-noise (although, most likely slower-switching) RF signal source will add flexibility to the tester.

References

[1] Pozar, D. M., *Microwave Engineering,* Boston: Addison-Wesley, 1993, pp. 582–594.

[2] Witte, R. A., *Spectrum and Network Measurements,* Upper Saddle River, NJ: Prentice Hall.

[3] Johnson, J. B., "Thermal Agitation of Electricity in Conductors," *Physical Review,* Vol. 32, 1928, p. 97.

[4] Nyquist, H., "Thermal Agitation of Electricity in Conductors," *Physical Review* 32, 1928, p. 110.

[5] Schottky, W., "Small-Shot Effect and Flicker Effect," *Physical Review* 28, 1926, p. 74.

[6] Hewlett Packard, "Fundamentals of RF and Microwave Noise Figure Measurements," Application note 57-1, 1983.

[7] Friis, H. T., "Noise Figures of Radio Receivers," *Proceedings of the IRE,* July 1944, pp. 419–422.

[8] Hewlett Packard, "Noise Figure Measurement Accuracy: The Y-Factor Method," Application note 57-2, 1992.

[9] Lance, A. L., Seal, W. D., and Bayuk, F. J., "Noise Measurement Uncertainty," *Journal of Applied Measurements* Vol. 2, 1974, pp. 70–75.

[10] Boyd, D., "Calculate the Uncertainty of NF Measurements," *Microwaves & RF*, October 1999, pp. 93–102.

[11] Hewlett Packard, "10 Hints for Making Successful Noise Figure Measurements," Application note 57-3 (2002).

[12] IEEE Standard 1139, 1988.

[13] Ferre-Pikal, E. S., et al., "Draft Revision of IEEE Standard 1139—1988 Standard Definitions of Physical Quantities for Fundamental Frequency and Time Metrology—Random Instabilities," *1997 International Frequency Control Symposium Proceedings*, 1997.

Appendix A: Power and Voltage Conversions

The unit dBW is decibels relative to 1 watt. To obtain a dBW value from a value of power in watts, use the following:

$$P_{dBW} = 10 \log_{10}\left(\frac{P_W}{1W}\right) \tag{A.1}$$

The unit dBm is decibels relative to 1 milliwatt. To obtain a dBm value from a value of power in watts, use the following:

$$P_{dBm} = 10 \log_{10}\left(\frac{P_{mW}}{1\ mW}\right) \tag{A.2}$$

To obtain power in milliwatts from a power level specified in dBm, use the following:

$$P_{mW} = 10^{\left(\frac{P_{dBm}}{10}\right)} \tag{A.3}$$

Note that (A.1–A.3) are independent of characteristic impedance (Z_0); hence, they will work for any impedance.

If P_W is broken down to its constituents, then

$$P_W = \frac{V^2}{Z_0} \tag{A.4}$$

Placing (A.4) into (A.3) arrives at the relationship between voltage and dBm. Note that it is dependent on impedance (Z_0):

$$V = \sqrt{Z_0 (0.001) 10^{\left(\frac{P_{dBm}}{10}\right)}} \tag{A.5}$$

Often, for cable TV applications, an impedance-dependent unit called V_{dBmV} is used. It is defined as

$$V_{dBmV} = 20 \log_{10}\left(\frac{V_{mV}}{1\ mW}\right) \tag{A.6}$$

Note the "20" multiplier. This is due to the fact that it is a voltage ratio and that the decibel concept has been originally defined for power. Since power is defined as V^2/R, the V^2 term gives rise to the logarithmic "20" multiplier [$10 \log(X^2) = 20 \log(X)$].

$$V_{dB\mu V} = 20 \log_{10}\left(\frac{V_{\mu V}}{1\ \mu V}\right) \tag{A.7}$$

Substituting (A.5) into (A.6) yields the following relationship:

$$V_{dBmV} = 10 \log\left(\frac{Z_0}{0.001}\right) + P_{dBm} \tag{A.8}$$

For a 50-Ω device or circuit,

$$V_{dBmV} = 46.99 + P_{dBm} \tag{A.9}$$

For a 75-Ω device or circuit,

$$V_{dBmV} = 48.75 + P_{dBm} \tag{A.10}$$

Tables A.1 and A.2 are a means to demonstrate the relationships of the various power and voltage values. Note that the relationships between power in dBm and in watts are the same regardless of impedance. For example, in Table A.1, where

Table A.1 Relationship Between Power and Voltages in Linear and Logarithmic Scales at 50-Ohm Characteristic Impedances

P (dBm)	P (W)	V (dBmV)	V (V)
−100	1.0×10^{-13}	−53.01	2.0×10^{-6}
−50	1.0×10^{-8}	−3.01	0.0007
−40	1.0×10^{-7}	6.99	0.002
−30	1.0×10^{-6}	16.99	0.007
−20	1.0×10^{-5}	26.99	0.022
−10	0.0001	36.99	0.071
−5	0.00032	41.99	0.126
−4	0.00040	42.99	0.141
−3	0.00050	43.99	0.158
−2	0.00063	44.99	0.178
−1	0.00079	45.99	0.199
+0	0.001	46.99	0.224
+1	0.0013	47.99	0.251
+2	0.0016	48.99	0.282
+3	0.0020	49.99	0.316
+4	0.0025	50.99	0.354
+5	0.0032	51.99	0.398
+10	0.01	56.99	0.707
+20	0.1	66.99	2.236
+30	1	76.99	7.071
+40	10	86.99	22.36
+50	100	96.99	70.71
+100	1.0×10^{7}	146.99	22,361

Table A.2 Relationship Between Power and Voltages in Linear and Logarithmic Scales at 75-Ohm Characteristic Impedances

P (dBm)	P (W)	V (dBmV)	V (V)
−100	1.0×10^{-13}	−51.25	3.0×10^{-6}
−50	1.0×10^{-8}	−1.25	0.0009
−40	1.0×10^{-7}	8.75	0.003
−30	1.0×10^{-6}	18.75	0.009
−20	1.0×10^{-5}	28.75	0.027
−10	0.0001	38.75	0.087
−5	0.00032	43.75	0.154
−4	0.00040	44.75	0.173
−3	0.00050	45.75	0.194
−2	0.00063	46.75	0.218
−1	0.00079	47.75	0.244
+0	0.001	48.75	0.274
+1	0.0013	49.75	0.307
+2	0.0016	50.75	0.345
+3	0.0020	51.75	0.387
+4	0.0025	52.75	0.434
+5	0.0032	53.75	0.487
+10	0.01	58.75	0.866
+20	0.1	68.75	2.739
+30	1	78.75	8.660
+40	10	88.75	27.39
+50	100	98.75	86.60
+100	1.0×10^{7}	148.75	27,386

$Z_0 = 50\Omega$, −10 dBm corresponds to 0.1 mW. Referring to Table A.1, where $Z_0 = 75\Omega$, −10 dBm also corresponds to 0.1 mW. The differences between values in these two tables become apparent when Z_0 is considered, as in voltage or power in units of dBmV. In a 50-Ω environment, (Table A.1), −10 dBm corresponds to 71 mV, while in a 750-Ω environment, (Table A.2), −10 dBm corresponds to 87 mV. A common reference point for every engineer should be to note that 0 dBm is equivalent to 1 mW. Keeping this in mind will be handy for those back-of-the-envelope calculations.

Appendix B: RF Coaxial Connectors

RF coaxial connectors are the most important element in a cable system. These connectors will be used internally to the test equipment in addition to on the load board. High-quality coaxial cables have the potential to deliver all the performance a system requires, but they are often limited by the performance of the connectors. Impairments such as power loss, electrical noise, and intermodulation distortion, a major concern in today's communications systems, are minimized by the design and manufacturing techniques of these connectors.

Connectors generally come in both "male" and "female" sections. Higher-quality RF connectors are even sometimes designed and manufactured in male-female pairs to gain optimal performance. The selection of quality connectors is a critical area to achieving the necessary performance.

Often, quality connectors for production testing are well worth the extra money spent. Through repeated connecting and disconnecting of, for example, a load board, the connectors can exhibit mechanical wear (i.e., gold plating is removed and electrical properties change). To minimize potential problems, it is imperative to maintain clean connectors in a test system (using a lint-free swab and rubbing alcohol, for example) and to ensure that any nuts on the connectors are tightened to the proper torque specification provided by the manufacturer [1].

When microwave cables and connectors are used in test and measurement applications, their service life may be considerably reduced as a result of frequent use [2].

This appendix will provide a descriptive overview of commonly used connectors in RF and SOC production testing.

B.1 Type BNC Connector

The type BNC (Bayonet Neill Concelman) connector is a relatively low-frequency (dc to 4 GHz) general-purpose RF connector designed for use in 50-Ω and 75-Ω systems. Developed in the late 1940s as a miniature version of the type C connector, BNC is named after Amphenol engineer Carl Concelman. The BNC is a miniature quick-connect/-disconnect RF connector. It features two bayonet lugs on the female connector; mating is achieved with only a quarter turn of the coupling nut. BNC connectors usually have nickel-plated brass bodies, Teflon insulators, and either gold- or silver-plated center contacts. These low-cost connectors are typically available with die cast and molded components. The diameter of the male pin differs for

50- and 75-ohm versions and sometimes the 75-ohm version has markings. Because of the different pin diameters, damage may occur if these are inadvertently mixed.

B.2 Type C Connector

The type C (Concelman) connector is medium sized and weatherproof and designed to work up to 11 GHz in 50-Ω systems. The coupling is a two-stud bayonet lock. C connectors provide constant 50-Ω impedance, but may be used with 75-Ω cable, at lower frequencies (below 300 MHz) where no serious mismatch is introduced.

B.3 Type N Connector

The type N (Neill) connector is for use at up to 11 GHz in a 50-ohm environment. Named after Paul Neill of Bell Labs after its development in the 1940s, the N connector offered the first true microwave performance.

B.4 Type SMA Connector

The type SMA (subminiature version A) connector is one of the most commonly used connectors in RF and SOC test equipment. The SMA connector was developed in the 1960s. It uses a threaded interface. These connectors are for use in a 50-Ω environment and provide excellent electrical performance up to 18 GHz. SMA connectors are available in both standard and reverse polarities. Reverse polarity is a keying system accomplished with a reverse interface and ensures that reverse polarity interface connectors do not mate with standard interface connectors.

Since SMA connectors are commonly used in production-testing equipment, it is worthwhile to point out a very useful tip. When mating a male SMA connector (the one with the pin in the center) to a female, make sure to spin only the collar on the male connector. Engineers often (and incorrectly) spin the female connector. This has the effect of prematurely wearing the contact point mechanically where the male pin enters the female receiver, causing reduced performance due to debris from plating and lessened conductivity due to oxidation.

B.5 Type SMB Connector

The type SMB (subminiature B) connector is so named because it was the second subminiature design connector. Developed in the 1960s, the SMB is a smaller version of the SMA with snap-on coupling. It is designed for both 50-Ω and 75-Ω impedances and for operation up to 10 GHz.

B.6 Type SMC Connector

The type SMC (subminiature C) connector is so named because it was the third subminiature design connector. It has a threaded coupling with 10–32 threads. It is designed for both 50-Ω and 75-Ω impedances and for operation up to 10 GHz.

B.7 Type TNC Connector

The type TNC (Threaded Neill Concelman) connector is designed to operate at up to 11 GHz in a 50-Ω system. It was developed in the late 1950s and named after Amphenol engineer Carl Concelman. Designed as a threaded version of the BNC, the TNC series features screw threads for mating. TNC connectors are available in both standard and reverse polarity. Reverse polarity is a keying system accomplished with a reverse interface and ensures that reverse polarity interface connectors do not mate with standard interface connectors.

B.8 UHF Connector

The UHF type connector is designed to operate at up to only 300 MHz, but at any impedance. This is one of the oldest RF connectors and was developed in the 1930s by an Amphenol engineer named E. Clark Quackenbush. Invented for use in the radio industry, UHF is an acronym for ultra high frequency because, at the time, 300 MHz was considered high frequency [3]. UHF connectors have a threaded coupling. In the 1970s, a miniature version of the UHF connector that operates up to 2.5 GHz in a 50-Ω environment was introduced. Today, they are often used in mobile phones and in automotive systems or other places where size, weight, and cost factors are critical.

References

[1] Hewlett Packard, "Microwave Connector Care," manual part number 08510-90064, 1986.
[2] Huber-Suhner Corporation, 2003.
[3] Amphenol Corporation, 2003.

List of Acronyms and Abbreviations

1G	First generation
2.5G	2.5 generation
2G	Second generation
3G	Third generation
ac	Alternating current
ACIR	Adjacent channel interference ratio
ACLR	Adjacent channel leakage ratio
ACPR	Adjacent channel power ratio
ADC	Analog-to-digital converter
AGC	Automatic gain control
AM	Amplitude modulation
ARB	Arbitrary waveform generator
ATE	Automatic test equipment
AWG	Arbitrary waveform generator
BER	Bit error rate
BIST	Built-in self test
BPF	Bandpass filter
BPSK	Binary phase shift keying
BW	Bandwidth
CDMA	Code division multiple access
COO	Cost of ownership
COT	Cost of test
CSP	Chip scale package
CW	Continuous wave
DAC	Digital-to-analog converter
dB	Decibel
dc	Direct current
DDC	Direct downconversion
DECT	Digital Enhanced Cordless Telecommunications
DFT	Design for test
DFT	Discrete Fourier transform
DH	Data high
DIB	Device interface board

DM Data medium
DMM Digital multimeter
DNL Differential nonlinearity
DRAM Dynamic random access memory
DSB Double sideband
DSP Digital signal processing
DUT Device under test
EDGE Enhanced data for GSM evolution
ENOB Effective number of bits
ENR Excess noise ratio
ESD Electrostatic discharge
ETSI European Telecommunications Standards Institute
EVM Error vector magnitude
FCC Federal Communications Commission
FD Frequency division
FDMA Frequency Division Multiple Access
FFT Fast Fourier transform
FH Frequency hopping
FHSS Frequency-hopping spread spectrum
FM Frequency modulation
FPGA Field programmable gate array
FSK Frequency shift keying
FSR Full-scale range
GaAs Gallium arsenide
GFSK Gaussian frequency shift keying
GPRS General Packet Radio Service
GPS Global Positioning Satellite
GSM Global System for Mobile Communications
I Current
IBM International Business Machine
IC Integrated circuit
ICFT Initial carrier frequency tolerance
IDM Integrated device manufacturer
IF Intermediate frequency
IFFT Inverse fast Fourier transform
IFT Inverse Fourier transform
IL Insertion loss
IMD Intermodulation distortion
INL Integral nonlinearity
IP Intermodulation product
IP2 Second-order intermodulation product

IP3	Third-order intermodulation product
I/Q	In-phase, quadrature-phase
ISM	Industrial, scientific, medial
KGD	Known good die
LAN	Local area network
LB	Load board
LCC	Leadless chip carrier
LNA	Low noise amplifier
LO	Local oscillator
LPF	Lowpass filter
LSB	Least significant bit
LSI	Large-scale integration
MBPS	Mega bits per second
MC	Multicommunicator
MCM	Multichip module
MSB	Most significant bit
MSOP	Miniature small outline package
MTBF	Mean time between failures
MTTR	Mean time to repair
MUX	Multiplex or multiplexer
NF	Noise figure
NIST	National Institute of Standards and Technology
OFDM	Orthogonal frequency division multiplexing
OTA	Overall timing accuracy
PA	Power amplifier
PAE	Power-added efficiency
PAN	Personal area network
PC	Personal computer
PDA	Personal digital assistant
PHD	Phase detector
PIB	Probe interface board
PLL	Phase locked loop
PN	Pseudonoise
PPM	Parts per million
PRBS	Pseudorandom bit sequence
PSD	Power spectral density
PSK	Phase shift keying
QAM	Quadrature amplitude modulation
QAM64	Quadrature amplitude modulation 64 levels
QPSK	Quadrature phase shift keying
R	Resistor

RBW Resolution bandwidth
RF Radio frequency
RFIC Radio frequency integrated circuit
rms Root mean square
RSSI Received signal strength indicator
RX Receive
SCM Subcontract manufacturer
SI International system of units
Si Silicon
SIG Special interest group
SiGe Silicon germanium
SINAD Signal-to-noise and distortion
SNR Signal-to-noise ratio
SOC System on a chip
SOIC Small outline integrated circuit
SSB Single sideband
TD Time division
THD Total harmonic distortion
TOI Third-order intercept
TSOP Thin small outline package
TX Transmit
ULSI Ultra-large-scale integration
UMTS Universal Mobile Telephone System
UPH Units per hour
UUT Unit under test
V Voltage
VAGC Voltage automatic gain control
VCO Voltage controlled oscillator
VGA Variable gain amplifier
VLSI Very large-scale integration
VMU Voltage measuring unit
V_{RMS} Root-mean-squared voltage
VSWR Voltage standing wave ratio
W Watt
WCDMA Wideband CDMA
WLAN Wireless local area network
XOR Exclusive OR
ZIF Zero intermediate frequency

List of Numerical Prefixes

Prefix	Symbol	Factor	
exa	E	$\times 10^{18}$	(quintillion)
peta	P	$\times 10^{15}$	(quadrillion)
tera	T	$\times 10^{12}$	(trillion)
giga	G	$\times 10^{9}$	(billion)
mega	M	$\times 10^{6}$	(million)
kilo	k	$\times 10^{3}$	(thousand)
hecto	h	$\times 10^{2}$	(hundred)
deka	da	$\times 10^{1}$	(ten)
deci	d	$\times 10^{-1}$	(tenth)
centi	c	$\times 10^{-2}$	(hundredth)
milli	m	$\times 10^{-3}$	(thousandth)
micro	μ	$\times 10^{-6}$	(millionth)
nano	n	$\times 10^{-9}$	(billionth)
pico	p	$\times 10^{-12}$	(trillionth)
femto	f	$\times 10^{-15}$	(quadrillionth)
atto	a	$\times 10^{-18}$	(quintillionth)

About the Authors

Keith B. Schaub was born in 1971 in Houston, Texas. Graduating with honors from Willis High School in 1989 and receiving his B.S. in electrical engineering in 1993 from Texas A&M University, he worked at Texas Instruments Defense Systems and Electronics Group for 3 years as an RF microwave engineer. In 1996, he began working at Hewlett-Packard as an RF applications engineer developing automatic test equipment (ATE). In 1997, he received his M.S. in electrical engineering from the University of Texas, Dallas, and promptly transferred to a foreign assignment with Agilent Technologies in Germany as a senior applications engineer in the Wireless Center of Expertise. He is currently a senior consultant for Agilent's Wireless Center of Expertise. He has delivered numerous presentations, including the published "Concurrent-Parallel Testing of Bluetooth/802.11x Chip Sets," at SEMICON Europe 2003. He currently lives and works in Austin, Texas. Mr. Schaub can be reached at keith_schaub@yahoo.com.

Joe Kelly was born in 1970 in Flemington, New Jersey. He received a B.S. in electrical engineering, an M.S. in ceramic engineering, and a Ph.D. in ceramic and materials engineering from Rutgers University. His graduate work focused on the processing and electromechanical properties of high dielectric constant piezoelectric ceramics and modeling of loss mechanisms in ceramic resonators. His graduate work also consisted of an internship at the Army ResearchLaboratory, Fort Monmouth, New Jersey, and the design and implementation of numerous electrical and physical characterization tests of electroacoustic and piezoelectric materials. He has worked for Siemens, now Epcos, as a surface acoustic wave (SAW) filter design engineer. Dr. Kelly has spent the past 4 years in RF and mixed-signal semiconductor testing at Hewlett-Packard, now Agilent Technologies. He is currently a senior consultant for Agilent's Wireless Center of Expertise. He can be reached at joe_kelly1970@yahoo.com.

Edwin Lowery III was born in 1969 in Los Angeles, California. He spent his formative years in Houston, Texas, and at age 11 moved to San Jose, Costa Rica, where he lived for 6 years and completed his high school education. He received a BSEE with high honors from the Florida Institute of Technology in 1992. As an undergraduate he worked as an electrical engineering intern for IBM, McDonnell Douglas, and Harris Space Systems. In 1993, he received an MSEE from the Georgia Institute of Technology, specializing in digital design and DSP. In graduate school he was a research assistant for the Georgia Tech Research Institute Acoustics Lab. He has worked professionally in many different electrical engineering positions, including digital R&D, radio transceiver testing, and SOC product engineering. His first job was working for Motorola in RF systems design, and later he was a product engineer for the semiconductor products sector. He has spent the past 5 years in RF and mixed-signal semiconductor testing working for Hewlett-Packard, now Agilent Technologies. He is currently a senior consultant for Agilent's Wireless Center of Expertise. He has published several articles, including "Impedance Matching Techniques for RFIC Test" at the HP users group in 1999, "Bluetooth Testing with the 94000BT" at Agilent's Test Fest 2000, and "Integrated Cellular Transceivers: Challenging Traditional Test Philosophies" at Semicon West 2003. He currently lives in Austin, Texas, with his wife, Elizabeth, and son, Edwin IV.

Ashish Desai was born in 1978 in Columbus, Ohio. He received a B.S. in electrical engineering and biomedical engineering from Duke University in 2000. During his undergraduate career, he worked as an intern in the Digital Signal Processing Group at Motorola, as well as as a research assistant in the Biomedical Engineering Department at Duke. Since 2000, he has been working at Agilent Technologies in the field of ATE, doing work including DSP, embedded design, software design, and development of RF measurement algorithms. While working at Agilent, he has been pursuing his MSEE from Stanford University and will graduate in 2004.

Index

1/f noise, 197–98
   defined, 197
   measurements, 198
   PSD, 198

A

Accuracy, 42–45
   defined, 10
   impact on yield, 43
Adaptive power control, 102
Adjacent channel interference ratio (ACIR), 79
Adjacent channel interference tests, 133–34
   condition illustration, 134
   defined, 133
   filter specifications and, 134
   See also BER tests
Adjacent channel power ratio (ACPR), 79–82
   defined, 79
   measuring, 81–82
   as modulated power-out measurement, 81
   plot, 81
Amplifiers
   block diagram, 58
   cascaded, 59
   clipping of, 70
   RF SOC, 60
   two-port, 88–89
Analog-to-digital converters (ADCs), 22, 25, 139
Antialiasing
   filter, 155
   illustrated, 154
Arbitrary waveform generators (AWGs)
   dynamic performance of, 167
   INL/DNL for, 164–65
   SNR for, 159–60
Autoloaders, 4–5
Automated test equipment, *xvii*
   accuracy, 42–45
   calibration, 9
   configurations, 2
   defined, 3
   fixed costs, 39
   lifetime, 40
   recurring costs, 39–40
   utilization, 40–41
Automatic gain control (AGC), 14, 20–22
   block diagram, 21
   programming, 22
Automatic gain control flatness, 65–67
   example pseudocode, 66–67
   ideal, 65
   key parameters, 66
   nonideal, 65
Average power, 55–56
   defined, 54, 55
   equation, 56
   See also Power
Averaging, 212–13

B

Bandwidth
   ideal Bluetooth plot, 116
   jitter relationship, 213
   narrow, 107
   nonideal Bluetooth plot, 116
   resolution (RBW), 220, 221
BER tests, 132–37
   adjacent channel interference, 133–34
   blocking, 135
   carrier-to-interference, 133
   cochannel interference, 133
   inband and out-of-band blocking, 135
   intermodulation interference, 135–37
   maximum input power level, 137
   sensitivity, 132–33
Bit error rate (BER), 22, 96
   defined, 125
   FPGA setup, 129
   measurement with digitizer, 130–32
   method comparison, 131
   methods, 127

Bit error rate (BER) (continued)
  receiver measurements, 132–37
  receiver test, 125–27
  setup using programmable delay line, 127
  testing with digital pin, 128–30
  test setup block diagram, 126
  See also BER tests
Blocking BER tests, 135
Bluetooth
  data rates, 100–102
  defined, 97–98
  ideal bandwidth plot, 116
  introduction, 98–99
  modem block diagram, 126
  modulation, 100
  nonideal bandwidth plot, 116
  operation band, 98
  origins, 97–98
  packets, 100–102
  PLL, 103–4
  radio modem block diagram, 98
  radio parts, 102–3
  transmit spectrum, 115
BNC connectors, 229
Bridges characteristics, 89
Built-in self-test (BIST), 11–12

C

Calibration, 9
Carrier drift, 119–20
Carrier frequency drift, 119–20
C connectors, 230
Charge pumps, 104
Chip-scale packages (CSPs), 6
C/I tests, 133
Coaxial connectors, 229–31
Cochannel interference BER tests, 133
  condition illustration, 134
  defined, 133
  See also BER tests
Code division multiple access (CDMA), 79–81
  chip rate, 80
  defined, 79–80
Cold noise method, 209
Comb frequencies, 83
Complex FFTs, 168–71
  with amp imbalance, 170
  amplitude/phase balance with, 169–71
  complex time domain and, 169
  with phase imbalance, 171
Concurrent testing, 189–90
  defined, 189

uses, 189
Connectors, 229–31
  BNC, 229
  C, 230
  male/female sections, 229
  N, 230
  SMA, 230
  SMB, 230
  SMC, 230
  TNC, 231
  UHF, 231
  use of, 229
Contactors, 5–6
  choosing, 6
  cost-accuracy trade-offs, 6
  defined, 5
  technologies, 5
Convolution, 151
Correlation, 11
Cost
  fixed, 39
  recurring, 39–40
Cost of ownership (COO)
  defined, 33
  standard, 39
  See also Cost of test (COT)
Cost of test (COT), xv–xvi, 33–46
  accuracy, 42–45
  factors influencing, 45–46
  fixed cost, 39
  key parameters expression, 39
  lifetime, 40
  modeling parameters, 38–45
  multisite testing and, 45–46
  parallel testing and, 45–46
  pseudoparallel RF testing, 188–89
  recurring cost, 39–40
  SOC, paradigm shift, 37–38
  test engineer skill and, 46
  true parallel RF testing, 187–88
  utilization, 40–41
  water processing and, 33–36
  yield, 41–42
Couplers
  characteristics of, 90–91
  ideal properties, 90
  real, 90–91
  vector calibration, 91
Crest factor, 56–57

D

dBc, 166

Index

dBm, 166
dBV, 166
DC offsets, 146
   compensation, 164
   defined, 163–64
Decibels, 165
   defined, 53
   use of, 53
Demodulators, 23–24
   block diagram, 23
   defined, 23
   single-ended, 23
   See also Modulators; SOC devices
Design for testing (DFT), 11, 190–91
   defined, 190
   RF concepts, 191
Device interface boards (DIBs), 61
Devices under test (DUTs), 1
   interface board (DIB), 131
   internal noise, 199
   noise power, 199
   overtesting, 2
   wavelength effect on, 50
Differential nonlinearity (DNL), 165
Differential phase
   calculating, 110
   time vs., 110–12
Digital signal processing (DSP)
   functions, 25
   multithreading, 186–87
Digital-to-analog converters (DACs), 22
Digitizers
   BER measurement with, 130–32
   INL/DNL for, 164–65
   sampling rate, 130
   SNR for, 159–60
Discrete Fourier transform (DFT), 149–50
   defined, 149
   inverse, 149
   in real-time computing systems, 150
   See also Fourier transforms
Distortion, 72–79
   harmonic, 73–75
   intermodulation, 75–79
   receiver architecture considerations, 79
Dividers, 104
Double-sideband (DSB) mixers, 210
Drift
   carrier, 119–20
   frequency, 216
   frequency determination, 122–24
   synthesizer settling time and, 121

   VCO, 120
Dynamic measurements, 156–63
   coherent sampling, 156–59
   SINAD, 160–63
   SNR, 159

E

Effective noise temperature, 203
Effective number of bits (ENOB), 167
Equipment error, 211
Error vector magnitude (EVM), 22, 96, 137–43
   average, 141
   comparison, 142
   computation, 140
   defined, 138–39
   introduction, 137–43
   measurement, making, 139–41
   for modulation formats, 141
   peak, 141
   related signal quality measurements, 141
   tester block diagrams, 140
   use for production testing, 142–43
Excess noise ratio (ENR), 203–4
   defined, 203
   logarithmic values, 203–4
   noise sources, 204
Exclusive or (XOR), 127–28

F

Fast Fourier transforms (FFTs), 81
   complex, 168–71
   execution, 117
   number of, 116
Field programmable gate array (FPGA), 128
Filter testing, 82–84
   with comb frequencies, 83
   as power-out test, 82
Fixed cost, 39
Flicker noise, 197–98
   defined, 197
   measurements, 198
   PSD, 198
   ZIF architecture and, 27
   See also Noise
Fourier series, 147
Fourier transforms, 147–49
   commonly used, 148
   discrete (DFT), 149–50
   fast (FFTs), 81, 116, 117, 168–71
   inverse, 147

Frequency
  bin, 157
  drift, 216
  negative, 150–51
  pulling/pushing, 120–24
  translating devices, 209–10
Frequency domain
  relationship example, 153
  SINAD calculation in, 160
  time domain relationships, 152
  transformations, 152–54
Frequency-hopping spread spectrum (FHSS), 99–100

G

Gain, 58–61
  defined, 58
  measurements of wireless SOC devices, 60–61
  stages, 59
  total, 59
  as vector quantity, 59
Gain flatness, 61–67
  AGC, 65–67
  determining, 62
  illustrated, 62
  measurement illustration, 63
  measuring, 62, 63–65
  multitone for, 64
  ripple, 63
Gravity feed handlers, 3–4
Guard banding, 42

H

Handlers, 3–5
  defined, 3
  gravity feed, 3–4
  number of sites and, 4
  pick-and-place, 4
  size, 4
  types of, 3
Harmonic distortion, 73–75
  defined, 73
  measurement, 74
  specification, 73
  total (THD), 74
  See also Distortion

I

Image-rejection mixers, 18
Impulse transformations, 149
Inband blocking BER test, 135
Index time, 4
Industrial, scientific, and medical (ISM) band, 95, 99
  short-range devices, 99
  unlicensing and, 99
Initial carrier frequency tolerance (ICFT), 118–19
  defined, 119
  test, 119
In-phase, quadrature phase. *See* I/Q
Insertion loss, 59
Integral nonlinearity, 165
Integrated device manufacturers (IDMs), *xvi*, 37
Interleaving technique, 185–86
  defined, 185
  flow chart, 186
Intermodulation distortion (IMD), 75–79
  intercept graph, 77
  plot, 77
  receiver architecture considerations, 79
  third-order, 75
  See also Distortion
Intermodulation interference tests, 135–37
  condition illustration, 136
  defined, 135
  setup, 135
  See also BER tests
I/Q
  accurate, characterization, 168–71
  diagrams, 137–38
  digital modulation constellation, 214
  modulation, 168–71

J

Jitter, 168
  bandwidth relationship, 213
  phase, 216–17

L

Lifetime, 40
Load boards, 5
Loop filters, 29, 104
Low noise amplifiers (LNAs), 15
Lumped-element analysis, 50–51

M

Maximum input power level BER test, 137
Mean time between failures (MTBF), 41
Mean time to repair (MTTR), 41

Mismatch error, 211–12
Mixed-signal devices, 175
Mixer conversion compression, 72
Mixers, 16–19
   block diagram, 15
   defined, 16
   double-balanced, 18
   double sideband (DSB), 210
   as downconverter, 17, 18
   image-rejection, 18
   parameters, 17
   single-ended, 17
   single sideband (SSB), 210
   types of, 17
Modulated power, 54, 56–57
   crest factor, 56–57
   determining, 57
   See also Power
Modulation
   Bluetooth, 100
   I/Q, 168–71
   transmitter characteristics, 117–18
Modulators, 22–23
   block diagram, 23
   mixers, 22
   phase/amplitude distortion, 22
   phase splitter, 22
   See also SOC devices
Multisite testing, 45–46
Multithreading, 186–87
Multitone stimulus test setup, 83

N

N connectors, 230
Negative frequency, 150–51
Noise
   1/F, 197–98
   differences, 194–95
   introduction to, 193–99
   measurements, 193–222
   output power vs. temperature, 206
   phase, 213–22
   plasma, 198
   power density, 201–2
   quantization, 198
   quantum, 198
   shot, 197
   sources, 202
   temperature, 202–3
   thermal, 195–97
   types of, 194–98
Noise figure, 199–213
   defined, 199–200
   direct calculation of, 205
   equipment error and, 211
   external interfering signals and, 212
   mathematically calculating, 204–5
   measurements, 201, 205–9
   measurements, error calculation, 210–11
   measurements, with cold noise method, 209
   measurements, with Y-factor method, 205–8
   measurements on frequency translating devices, 209–10
   mismatch error and, 211–12
   production-test fixturing and, 212
Noise floor, 125, 198–99
   defined, 198
   effects on receiver, 219
   illustrated, 199
Nyquist sampling theory, 154–56

O

Ohm's Law, 49, 50
Organization, this book, *xvi–xvii*
Out-of-band blocking BER test, 135

P

Parallel testing, 45–46
   alternative methods, 182–84
   of digital/mixed-signal devices, 175
   illustrated, 176
   pseudoparallel, 180–82
   pseudoparallel, COT advantages/disadvantages, 188–89
   quad-site Bluetooth setup, 183
   of RF devices, 175–78
   of SOC devices, 178–79
   true, 179–80
   true, COT advantages/disadvantages, 187–88
Peak-to-peak input voltages, 146
Phase detectors (PHDs), 104
Phase jitter, 216–17
Phase locked loops (PLLs), 28–30
   blocks, 103–4
   Bluetooth, 103–4
   charge pumps, 104
   components, 28, 29
   defined, 28
   divider, 104
   functioning of, 104–5
   LPF, 104

Phase locked loops (PLLs) (continued)
    PHD, 104
    VCO, 104
Phase noise, 168, 213–22
    defined, 213, 214
    of fast-switching RF signal sources, 222
    high power measurement, 217
    illustrated, 219
    introduction, 213–14
    low power measurement, 217
    measurement example, 220–22
    measurements, making, 218–19
    measurements, with spectrum analyzer, 220
    measurement trade-offs, 217–18
    specification, 213
    spectral density-based definition, 216
    thermal effects on, 217
Pick-and-place handlers, 4
Plasma noise, 198
Power, 54–55
    absolute, 54
    average, 54, 55–56
    conversions, 225–27
    in dBm, 54
    frequency increase and, 52
    importance of, 52–53
    modulated, 54, 56–57
    pulse, 54, 56
    RMS, 57–58
    thermal noise, 197
    time vs., 55, 106–10
    total, 194
    voltage relationship to, 226, 227
Power-added efficiency (PAE), 67–68
    defined, 67
    higher, 68
Power amplifiers (PAs), 15–16
    block diagram, 15
    use of, 16
Power compression, 69–72
    algorithm, 71–72
    defined, 69
    graph, 71
Power measurements
    history of, 51–52
    inconsistencies, 52
    units and definitions, 53
    voltage measurements vs., 49–50
Power spectral density (PSD), 193–94
    defined, 193–94
    illustrated, 195
    units, 194

Prober interference board (PIB), 8
Production noise measurements, 193–222
Production testing
    characterization vs., 1–2
    contactor sockets, 5–6
    equipment, 2, 4–9
    EVM use for, 142–43
    fixturing, 212
    introduction, 1–12
    moving beyond, 175–92
    multisite, 45–46
    parallel, 45–46
    RF device, 30, 49–92
    SOC device, 30
    of SOC devices, 95–143
    wafer probing during, 8
Pseudoparallel RF testing, 180–82
    concurrent setup, 181
    COT advantages/disadvantages, 188–89
    downconverting architecture, 181
    setup, 181
Pulse power, 54
    defined, 56
    equation, 56
    See also Power

Q

Quadrature phase shift keying (QPSK), 138, 143
Quantization noise, 198
Quantum noise, 198

R

Rack-and-stack testers, 2–3
    calibration, 9
    uses, 2–3
Radio frequency. *See* RF devices
Received signal strength indicators (RSSIs), 14
Receivers, 24–25
    block diagram, 24
    block diagram of Bluetooth modem, 126
    defined, 24–25
    sensitivity, 25
    See also SOC devices; transmitters
Receiver tests, 124–32
    BER measurement with digitizer, 130–32
    BER methods, 127
    BER testing with digital pin, 128–30
    bit error rate, 125–27
    design, 125
    FPGA method, 128

Index 247

programmable delay line method, 127–28
Receive signal strength indicator (RSSI), 191
Recurring cost, 39–40
Reflection coefficient, 87
Repeatability, 10–11
Return loss, 87
RF devices, *xv*, 13–30
    block diagram representations, 15
    low noise amplifier (LNA), 15
    mixer, 16–18
    parallel testing of, 175–78
    power amplifier (PA), 15–16
    production testing, 30, 49–92
    switch, 19–20
    tests, 30
    transfer function for, 68–69
RF integrated circuit (RFIC) devices, 6
RF wafer probing, 6–9
RMS power, 57–58
    defined, 57
    equation, 58
    See also Power

S

Sampling
    basics and conventions, 145–46
    coherent, 156–59
    illustrated, 154
    jitter and, 168
    noncoherent, 158
    Nyquist, 154–56
    rate, 157
Scalar measurements, 86–88
Schottky noise, 197
Sensitivity BER tests, 132–33
Serial protocol interface (SPI)
    clock, 112
    three-wire, 112, 113
Shannon's theorem, 154
Shot noise, 197
Signal, noise, and distortion (SINAD), 74–75
    calculation, 160
    equation, 74
Signal-to-noise ratio (SNR), 137
    for AWGs, 159–60
    degradation, 200
    for digitizers, 159–60
    harmonics and, 160
Single-sideband (SSB) mixers, 210
SMA connectors, 230
SMB connectors, 230
SMC connectors, 230

SOC
    COT paradigm shift, 37–38
    digital control of, 112–13
    early testing and, 36–37
    flexibility, 41
    integration levels, 96–97
    manufacturers, 38
    RF-to-analog configuration, 96
    RF-to-digital configuration, 97
    RF-to-RF configuration, 96
    testing, 37
    wafer probing, 6–9
    wireless configurations, 96–97
SOC devices, *xv*, 13–30
    defined, 14
    demodulator, 23–24
    DFT in, 11
    modulator, 22–23
    parallel testing of, 178–79
    PLLs, 28–30
    production testing of, 95–143
    radios, 35
    receiver, 24–25
    RF/baseband, 14
    RF/digital, 15
    RF input/output, 14
    tests, 30
    transceiver, 25–26
    transmitter, 24
    VGA, 20–22
    wireless, gain measurements, 60–61
S-parameters, 84–91
    defined, 84–85
    of generic amplifier, 88
    introduction to, 84
    measurements, 85
    measurements, relationship of, 89
    realization, 89
    scalar measurements related to, 86–88
    specification, 85
    two-port device, 85–86
    two-port realization, 92
Spectrum analyzer
    phase noise measurement with, 220, 221
    resolution bandwidth (RBW), 220, 221
Static measurements, 163–65
    DC offset, 163–64
    INL/DNL, 164–65
Subcontract manufacturers (SCMs), *xvi*, 37
Superheterodyne wireless radio, 26
Switches, 19–20
    FET, 20

Switches (continued)
  GaAs, 19–20
  PIN, 19
Synthesizer settling time, 105–6
  defined, 105
  drift and, 121
  as key performance factor, 105
  measurement test setup, 107
  pulling and, 121
  result using power-vs.-time method, 109
  testing, 106
  worst-case, 108
System-on-a-chip. See SOC devices

T

Test cell, 10
Test engineers, skill, 46
Test equipment, 2
  calibration, 9
  handlers, 3–5
  interfacing, 3–9
  load boards, 5
Test floor, 10
Test houses, 10
Test programs, 2
Thermal noise, 195–97
  defined, 195
  equations, 196
  power, 197
  See also Noise
Time
  differential phase vs., 110–12
  power vs., 106–10
Time domain
  complex, 169
  frequency domain relationships, 152
  relationship examples, 153
  transformations, 152–54
TNC connectors, 231
Total power, 194
Transceivers, 25–26
  block diagram, 25
  common LO, 26
  defined, 25
  superheterodyne, 26
  ZIF, 27
Transfer function, 68–69, 88–89
Transformation formulas, 166
Transmission coefficient, 87
Transmission line theory, 50–51
Transmitters, 24
Transmitter tests, 113–24

  carrier frequency drift, 119–20
  frequency pulling/pushing, 120–24
  initial carrier frequency, 118–19
  modulation characteristics, 117–18
  output spectrum, 114–17
  parameters, 114
  VCO drift, 120
True RF testing, 179–80
  architecture illustration, 180
  COT advantages/disadvantages, 187–88
  defined, 180
  efficiency, 180
  See also Parallel testing
Tuned radio frequency (TRF), 26

U

UHF connectors, 231
Universal Mobile Telephone System (UMTS), 79
Unloaders, 4–5
Utilization, 40–41
  defined, 40
  equation, 41

V

Variable gain amplifiers (VGAs), 20–22
  defined, 20
  illustrated, 21
Voltage
  conversions, 225–27
  power relationship to, 226, 227
Voltage controlled oscillators (VCOs), 26, 28, 29
  drift, 120
  input/output, 104
  PLL, 104
Voltage standing wave ratio (VSWR), 19

W

Wafer probing, 6–9
  probe card, 7–8
  in production testing, 8
  RF measurements performed with, 7
  station, 8
Wafers
  chip volume and, 36
  dimensions, increasing, 34
Wafer processing, 33–36
Windows, 156–59
  Blackman, 158, 159
  Hamming, 158, 162

Hanning, 162
rectangular, 162, 163
types of, 157
Wireless radio
architectures, 26
superheterodyne, 26
ZIF, 26–28

Y

Y-factor, 204
measurement process, 207
measuring noise figure using, 205–8
noise-figure measurement setup, 206–7
Yield, 41–42
accuracy impact on, 43
as shared COT element, 44
tester accuracy vs., 42

Z

Zero-intermediate frequency (ZIF) transceivers, 15, 26–28, 172
defined, 26–27
flicker noise and, 27
illustrated, 27

Recent Titles in the Artech House Microwave Library

Advanced Techniques in RF Power Amplifier Design, Steve C. Cripps

Automated Smith Chart, Version 4.0: Software and User's Manual,
  Leonard M. Schwab

Behavioral Modeling of Nonlinear RF and Microwave Devices,
  Thomas R. Turlington

Computer-Aided Analysis of Nonlinear Microwave Circuits,
  Paulo J. C. Rodrigues

Design of FET Frequency Multipliers and Harmonic Oscillators, Edmar Camargo

Design of Linear RF Outphasing Power Amplifiers, Xuejun Zhang,
  Lawrence E. Larson, and Peter M. Asbeck

Design of RF and Microwave Amplifiers and Oscillators,
  Pieter L. D. Abrie

Distortion in RF Power Amplifiers, Joel Vuolevi and Timo Rahkonen

*EMPLAN: Electromagnetic Analysis of Printed Structures in Planarly Layered
  Media, Software and User's Manual,* Noyan Kinayman and M. I. Aksun

Feedforward Linear Power Amplifiers, Nick Pothecary

Generalized Filter Design by Computer Optimization,
  Djuradj Budimir

High-Linearity RF Amplifier Design, Peter B. Kenington

High-Speed Circuit Board Signal Integrity, Stephen C. Thierauf

Intermodulation Distortion in Microwave and Wireless Circuits,
  José Carlos Pedro and Nuno Borges Carvalho

Lumped Elements for RF and Microwave Circuits, Inder Bahl

Microwave Circuit Modeling Using Electromagnetic Field Simulation,
  Daniel G. Swanson, Jr. and Wolfgang J. R. Hoefer

Microwave Component Mechanics, Harri Eskelinen and
  Pekka Eskelinen

Microwave Engineers' Handbook, Two Volumes,
  Theodore Saad, editor

Microwave Filters, Impedance-Matching Networks, and Coupling Structures,
  George L. Matthaei, Leo Young, and E.M.T. Jones

Microwave Materials and Fabrication Techniques, Third Edition,
  Thomas S. Laverghetta

Microwave Mixers, Second Edition, Stephen A. Maas

Microwave Radio Transmission Design Guide, Trevor Manning

Microwaves and Wireless Simplified, Thomas S. Laverghetta

Neural Networks for RF and Microwave Design, Q. J. Zhang and K. C. Gupta

Nonlinear Microwave and RF Circuits, Second Edition, Stephen A. Maas

QMATCH: Lumped-Element Impedance Matching, Software and User's Guide, Pieter L. D. Abrie

Practical RF Circuit Design for Modern Wireless Systems, Volume I: Passive Circuits and Systems, Les Besser and Rowan Gilmore

Practical RF Circuit Design for Modern Wireless Systems, Volume II: Active Circuits and Systems, Rowan Gilmore and Les Besser

Production Testing of RF and System-on-a-Chip Devices for Wireless Communications, Keith B. Schaub and Joe Kelly

Radio Frequency Integrated Circuit Design, John Rogers and Calvin Plett

RF Design Guide: Systems, Circuits, and Equations, Peter Vizmuller

RF Measurements of Die and Packages, Scott A. Wartenberg

The RF and Microwave Circuit Design Handbook, Stephen A. Maas

RF and Microwave Coupled-Line Circuits, Rajesh Mongia, Inder Bahl, and Prakash Bhartia

RF and Microwave Oscillator Design, Michal Odyniec, editor

RF Power Amplifiers for Wireless Communications, Steve C. Cripps

RF Systems, Components, and Circuits Handbook, Ferril Losee

Stability Analysis of Nonlinear Microwave Circuits, Almudena Suárez and Raymond Quéré

TRAVIS 2.0: Transmission Line Visualization Software and User's Guide, Version 2.0, Robert G. Kaires and Barton T. Hickman

Understanding Microwave Heating Cavities, Tse V. Chow Ting Chan and Howard C. Reader

For further information on these and other Artech House titles,
including previously considered out-of-print books now available through our
In-Print-Forever® (IPF®) program, contact:

Artech House Publishers	Artech House Books
685 Canton Street	46 Gillingham Street
Norwood, MA 02062	London SW1V 1AH UK
Phone: 781-769-9750	Phone: +44 (0)20 7596 8750
Fax: 781-769-6334	Fax: +44 (0)20 7630 0166
e-mail: artech@artechhouse.com	e-mail: artech-uk@artechhouse.com

Find us on the World Wide Web at:
www.artechhouse.com